Urban Cegrell

Capacities in Complex Analysis

Aspects of Mathematics
Aspekte der Mathematik

Editor: Klas Diederich

All volumes of the series are listed on pages 154—155.

Urban Cegrell

Capacities in Complex Analysis

Springer Fachmedien Wiesbaden GmbH

CIP-Titelaufnahme der Deutschen Bibliothek

Cegrell, Urban:
Capacities in complex analysis / Urban Cegrell. —
Braunschweig; Wiesbaden: Vieweg, 1988
   (Aspects of mathematics: E; Vol. 14)
   ISBN 978-3-528-06335-1

NE: Aspects of mathematics / E

Prof. Dr. *Urban Cegrell*
Department of Mathematics, University of Umeå, Sweden

AMS Subject Classification: 32 F 05, 31 B 15, 30 C 85, 32 H 10, 35 J 60

Vieweg is a subsidiary company of the Bertelsmann Publishing Group.

Produced by Lengericher Handelsdruckerei, Lengerich

ISBN 978-3-528-06335-1   ISBN 978-3-663-14203-4 (eBook)
DOI 10.1007/978-3-663-14203-4

Contents

Introduction

The purpose of this book is to study plurisubharmonic and analytic functions in $\mathbb{C}^n$ using capacity theory. The case n=1 has been studied for a long time and is very well understood. The theory has been generalized to $\mathbb{R}^n$ and the results are in many cases similar to the situation in $\mathbb{C}$. However, these results are not so well adapted to complex analysis in several variables - they are more related to harmonic than pluriharmonic functions.

Capacities can be thought of as a non-linear generalization of measures; capacities are set functions and many of the capacities considered here can be obtained as envelopes of measures.

In the $\mathbb{R}^n$ theory, the link between functions and capacities is often the Laplace operator - the corresponding link in the $\mathbb{C}^n$ theory is the complex Monge-Ampère operator. This operator is non-linear (it is n-linear) while the Laplace operator is linear. This explains why the theories in $\mathbb{R}^n$ and $\mathbb{C}^n$ differ considerably. For example, the sum of two harmonic functions is harmonic, but it can happen that the sum of two plurisubharmonic functions has positive Monge-Ampère mass while each of the two functions has vanishing Monge-Ampère mass. To give an example of similarities and differences, consider the following statements. Assume first that Ω is an open subset of $\mathbb{R}^n$ and that K is a closed subset of Ω. Consider the following properties that K may or may not have.

(i) For every $z_0 \in K$ there is a subharmonic function φ on Ω such that

$$\overline{\lim_{z \to z_0}} \varphi(z) < \varphi(z_0), \quad \text{where} \quad z \in K, \ z \neq z_0.$$

(ii) There is a subharmonic function φ on Ω, $\varphi \not\equiv -\infty$, such that

$$K \subset \{z \in \Omega; \ \varphi(z) = -\infty\}.$$

(iii) There is a locally upper bounded family $(\varphi_i)_{i \in I}$ of subharmonic functions on Ω such that $K \subset \{z \in \Omega; \ \varphi(z) < \varphi^*(z)\}$, where $\varphi(z) = \sup_{i \in I} \varphi_i(z)$ and $\varphi^*(z) = \overline{\lim_{z' \to z}} \varphi(z')$.

(iv) If φ is subharmonic outside K and if

$$\overline{\lim_{z' \to z}} \varphi(z') < +\infty, \quad \text{where} \quad z' \notin K, \quad \forall z \in \Omega,$$

then φ extends to a uniquely determined subharmonic function on Ω.

In classical potential theory, it is a theorem that these properties are equivalent and the compact sets that have the properties are exactly those with vanishing Newton capacity.

To study the corresponding properties in $\mathbb{C}^n$, $\mathbb{R}^n$ has to be replaced by $\mathbb{C}^n$ and the subharmonic functions by the plurisubharmonic functions. Conditions (i)-(iv) are then transformed into conditions (i')-(iv') and they are no longer equivalent but:

$$(i') \underset{\not\Leftarrow}{\Rightarrow} (ii') \Longleftrightarrow (iii') \underset{\not\Leftarrow}{\Rightarrow} (iii) \underset{\not\Leftarrow}{\Rightarrow} (iv').$$

(cf. the last reference in Section X).

Section I and II are concerned with general capacity theory, Section III capacities related to function classes. In Section IV and V we specialize to subharmonic and pluri-subharmonic functions, respectively.

In Section V and VI we also study the complex Monge-Ampère operator. In Section VII, VIII and IV we use the results obtained to study certain plurisubharmonic functions, while Sections IX, X and XI are devoted to capacities and analytic functions. Finally, Section XII is concerned with the capacity generated by representing measures on the spectrum of the algebra of bounded analytic functions.

This book contains the notes I prepared in November 1983 for a couple of seminars at the University of Uppsala. These notes were made into a more complete form during a series of lectures at the University of Umeå in the fall 1985 and at Université Paul Sabatier, Toulouse in December 1986.

<u>General references</u>

Capacity theory

Gustave Choquet, Lectures on analysis, 1-3. W.A. Benjamin, 1969.

O.D. Kellog, Foundations of potential theory. Springer-Verlag, 1929.

N.S. Landkof, Foundations of modern potential theory. Springer-Verlag, 1972.

Complex analysis in several variables

L. Hörmander, An introduction to complex analysis in several varibles. North Holland, 1973.

Steven G. Krantz, Function theory of several complex variables. John Wiley & Sons, 1982.

P. Lelong, Plurisubharmonic functions and positive differential forms. Gordon and Breach, 1969.

List of Notations

Notation	Meaning

$\mathbb{N}$ the natural numbers

$\mathbb{R}$ the real numbers

$\mathbb{C}$ the complex numbers

$P(U)$ all the subsets of U

$\overline{U}^n$ the product space $U \times \ldots \times U$

χ_U the characteristic function of U

$C^p(\Omega)$ the class of real or complex valued functions on Ω with continuous derivatives or or order $\leq p$

∂ the differential operator $\Sigma \dfrac{\partial}{\partial z_j} dz_j$

$\overline{\partial}$ the differential operator $\Sigma \dfrac{\partial}{\partial \overline{z}_j} d\overline{z}_j$

∂U the boundary of the set U

the exterior product

$L^p(\mu, U)$ μ is a measure on U and $L^p(\mu, U)$ is the class of μ-measurable functions on U with $\displaystyle\int_U |f|^p d\mu < +\infty$

I Capacities

Definition. Let U be a σ-compact Hausdorff-space. A _capacity_ c on U is a set function defined on $P(U)$, the subsets of U, with the following properties:

i) $P(U) \ni E \mapsto c(E) \in \mathbb{R}^+$, $c(\emptyset) = 0$

ii) $P(U) \ni E_s \nearrow E$, $s \to +\infty$, $\Rightarrow \sup_{s \in \mathbb{N}} c(E_s) = c(E)$

iii) If K_s, $s \in \mathbb{N}$ is a decreasing sequence of compact subsets

of U, $K = \bigcap_{s=1}^{\infty} K_s$, then

$\inf_{s \in \mathbb{N}} c(K_s) = c(K)$.

Definition. A set function satisfying property i) and ii) above is called a _precapacity_.

Example I:1. If μ is a positive Radon measure then the outer measure μ^* is a capacity.

Capacities are thus a non-linear generalization of measures. Observe that no linearity is assumed e.g. if $f: \mathbb{R}^+ \to \mathbb{R}^+$ is a continuous and increasing function vanishing at the origin, then $f \circ c$ is a capacity for every capacity c.

Definition. Let μ_s, $s \in \mathbb{N}$ and μ be measures. We say that μ_s tends to μ weakly and write $\mu_s \rightharpoonup \mu$ if

$$\lim_{s \to \infty} \int \varphi d\mu_s = \int \varphi d\mu, \quad \forall \varphi \in C_0(U).$$

Lemma I:1. If $(\mu_s)_{s=1}^{\infty}$ is a sequence of positive measures such that $\mu_s \rightrightarrows \mu$ and if φ is an upper semicontinuous function with compact support then

$$\overline{\lim_{s \to \infty}} \int \varphi d\mu_s \leq \int \varphi d\mu.$$

Proof. Choose $\varphi_i \in C_0(U): \varphi_i \searrow \varphi$, $i \to +\infty$. Then

$$\overline{\lim_{s \to \infty}} \int \varphi d\mu_s \leq \overline{\lim_{s \to +\infty}} \int \varphi_j d\mu_s = \int \varphi_j d\mu, \quad \forall j \in \mathbb{N}.$$

By monotone convergence, $\lim\limits_{j \to +\infty} \int \varphi_j d\mu = \int \varphi d\mu.$

Theorem I:1. Let M be a weak*-compact set of positive measures. Then $c(E) = \sup\limits_{\mu \in M} \mu^*(E)$ is a capacity.

Proof. Everything but iii) is clear. Given $\varepsilon > 0$. For every K_j, choose $\mu_j \in M$ with $\mu_j(K_j) \geq c(K_j) - \varepsilon$. Since M is compact, there is an accumulation point $\mu \in M$ for $(\mu_j)_{j=1}^{\infty}$. There is a continuous function $\varphi > \chi_K$ so that $\int (\varphi - \chi_K) d\mu < \varepsilon$ and since

$$\{\nu \in M; \ |\int \varphi d\mu - \int \varphi d\nu| < \varepsilon\}$$

is a weak*-neighborhood of μ, there is a $j \in \mathbb{N}$ so that $\chi_{K_j} \leq \varphi$ and

$$|\int \varphi d\mu - \int \varphi d\mu_j| < \varepsilon.$$

Therefore, $c(K) \leq c(K_j) \leq \int \varphi d\mu_j + \varepsilon \leq \int \varphi d\mu + 2\varepsilon \leq \mu(K) + 3\varepsilon$ which proves the theorem.

Corollary I:1. If $0 < \varphi$ is a lower semicontinuous function on $U \subset R^n$, U compact, and if $M = \{\mu \geq 0; \ 0 \leq \varphi * d\mu \leq 1\}$ then

$\sup\limits_{\mu \in M} \mu^*(E)$ is a capacity. ($\varphi * d\mu$ stands for convolution.)

Proof. It is clear from Lemma I:1 that M is weak*-compact.

II Capacitability

Let c be a capacity.

Definition. A set E is said to be c-<u>capacitable</u> if

$$c(E) = \sup\{c(K); E \supset K, \text{ compact}\}.$$

Definition. A set E is said to be <u>universally capacitable</u> if E is c-capacitable for every capacity c on U.

Before proceeding to the main theorem of this section we need some topological concepts.

U   is as always a σ-compact Hausdorff space.

G_δ consists of the sets that can be written as a denumerable intersection of open sets.

F_σ consists of the sets that can be written as a denumerable union of compact sets.

$K_{\sigma\delta}$ consists of the sets that can be written as a denumerable intersection of sets from F_σ.

A set E in a Hausdorff space is called <u>K-analytic</u> if there is a $K_{\sigma\delta}$-set F in a compact space W and a continuous map $f:F \to E$ such that $E = f(F)$. A complete, separable and metric space is called <u>Polish</u>.

Finally, a continuous image of a Polish space is called an <u>analytic set.</u>

Proposition II:1. If E is a K-analytic set in a compact space U, then there is a $K_{\sigma\delta}$-set F in a compact set W and a continuous map $f: W \to U$ with $f(F) = E$.

Proof. Put $\Gamma = \{(x, f(x)), x \in F\}$ where $f: F \to E$ and $W \supset F$ is as in the definition of K-analytic. Then $E = \mathrm{proj}\ \Gamma$ and $\Gamma = \overline{\Gamma} \cap (F \times U)$ (closure in $W \times U$) since Γ is closed in $F \times U$ by the continuity of f. Since $W \times U$ is compact so is $\overline{\Gamma}$ and since $F \times U$ is a $K_{\sigma\delta}$-set so is Γ $(=\overline{\Gamma} \cap F \times U)$. Thus $\mathrm{proj}: \overline{\Gamma} \to E$ is the required function.

Theorem II:1.

a) Every Hausdorff space that is a continuous image of K-analytic set is K-analytic.

b) Denumerable unions and intersections of K-analytic sets are K-analytic.

Theorem II:2

a) Every Polish space is homeomorphic to a G_{δ}-set contained in a compact metric space.

b) Every Borel set in a Polish space is K-analytic.

Proof (of Theorem II:1).

a) is clear because of the transitivity of continuity.

b) Let $F_n \supset B_n \xrightarrow{f} E$ where B_n is a $K_{\sigma\delta}$-set in the compact space F_n and $f_n(B_n) = X_n \subset E$. Consider $F_0 = \Sigma\ F_n$ and let F be a compact space containing F_0 ($\Sigma\ F_n$ denotes for a moment the topological sum $\Sigma\ F_n = \cup F_n \times \{n\}$). Let

$B = \Sigma B_n$ and define $f: B \to E$ by $f(x) = f_n(x)$ if $x \in B_n$. Then f is continuous with $f(B) = \cup X_n$. Furthermore $B_n = \cap_i B_{n,i}$ where $B_{n,i}$ is a K_σ of F_n, $B = \cup_n B_n \times \{n\} = \cup_n \cap_i B_{n,i} \times \{n\} = \cap_i (\cup_n B_{n,i} \times \{n\})$. Therefore B is a $K_{\sigma\delta}$ of F so $\cup_n X_n$ is K-analytic.

It remains to prove that $\cap_n X_n$ is K-analytic. Let $F = \Pi B_n$ the product space and let $C = \Pi B_n$. The set C is the intersection of the cylinders b_n of F where $b_n = B_n \times \Pi_{p \neq n} F_p$ where each b_n is $K_{\sigma\delta}$ hence so is C. Furthermore, we denote by f_n the canonical extension of f_n to b_n. Every f_n is then defined and continuous on C. Then for every pair i, j

$$\{z \in C: f_j(z) = f_i(z)\}$$

is a closed subset of C. Since C is $K_{\sigma\delta}$ it follows that

$$B = \cap_{i,j} \{z \in C: f_i(z) = f_j(z)\}$$

is a $K_{\sigma\delta}$-set in F.

Define now f to be the restriction to B of f_n. This restriction is continuous and since $f_n(B) \subset X_n$, $\forall n \in \mathbb{N}$ we have $f(B) \subset \cap X_n$. On the other hand, $\cap X_n \subset f(B)$. For let $y \in \cap X_n$, for every n there is an $x_n \in B_n$ such that $f_n(x_n) = y$. The point $(x_n) \in B$ and therefore $f(x_n) = y$. Therefore $A = f(B)$ which proves the theorem.

Proof (of Theorem II:2). a) Let $I = [0,1]$ and $I^{I\!N} = \underset{I\!N}{X}[0,1]$.

Then $I^{I\!N}$ is a compact, separable and metrizable space with

metric $D(a,b) = \sum\limits_{j=1}^{\infty} \dfrac{|a_j - b_j|}{2^j(1+|a_j - b_j|)}$. Let now X be a Polish

space with metric d. (We can assume that $0 \leq d \leq 1$). We

consider the map h,

$$X \ni x \overset{h}{\mapsto} (d(x,x_n))_{n=1}^{\infty} \in I^{I\!N}$$

where $(x_n)_{n=1}^{\infty}$ is dense in X. We define $Y = h(X)$ and claim

1) $X \overset{h}{\to} Y$ is a homeomorphism.

2) Y is a G_δ-set.

For if $y_i \to y$ in X, then $d(y_j,x_n) \to d(y,x_n)$, $j \to +\infty$, $\forall n \in I\!N$

so therefore $D(h(y_j)) \to D(h(y))$, $j \to +\infty$ which means that h is

continuous. Since $(x_n)_{n=1}^{\infty}$ is dense in X, h is one-to-one.

If $h(y_j) \to h(y) \in Y$, then $d(y_j,x_n) \to d(y,x_n)$, $j \to +\infty$, $\forall n \in I\!N$.

This means that $d(y_j,y) \to 0$, $j \to +\infty$ for given $\varepsilon > 0$, choose n_ε

so that $d(y,x_{n_\varepsilon}) < \varepsilon$. Then $0 \leq d(y_j,y) \leq d(y_j,x_{n_\varepsilon}) + d(y,x_{n_\varepsilon})$

so $\overline{\lim\limits_{j \to +\infty}} \, d(y_j,y) < \varepsilon$.

It remains to show that Y is a G_δ-subset of $I^{I\!N}$.

Let Q_n be the set of points z in $I^{I\!N}$ such that there

is an $r_n > 0$ such that

$D(h(x_1),z) < r_n$; $D(h(x_2),z) < r_n \Rightarrow d(x_1,x_2) < \dfrac{1}{n}$. Then each

Q_n is open.

Assume that $z_0 \in Y \cap \bigcap\limits_{n=1}^{\infty} Q_n$ and take $x^p \in X$ so that

$D(h(x^p), z_0) \to 0$, $p \to +\infty$. Given $\varepsilon > 0$. Take $h > \frac{1}{\varepsilon}$, choose the corresponding r_n and then p_n so that

$p \geq p_n \Rightarrow D(h(x^p), z_0) < r_n$. Hence, if $p, q \geq p_n$

$$d(x^p, x^q) < \frac{1}{n} < \varepsilon.$$

Therefore $(x^p)_{p=1}^{\infty}$ is a Cauchy sequence in X so $\lim_{p \to +\infty} x^p = x_0 \in X$ and therefore $h(x_0) = z_0$. Finally,

$$Y = \bigcap_{j=1}^{\infty} \; \bigcup_{z \in Y} \{y \in I^{I\!N}; \; D(y, z) < \tfrac{1}{j}\}.$$

b)   Since $I^{I\!N}$ is compact every open subset of $I^{I\!N}$ is a F_σ-set. Therefore, by a) every Polish space is K-analytic. Suppose now U is an open subset of a Polish space; we claim that U is Polish. It is clear that $I\!R \times E$ is Polish and if d is a metric on E,

$$V = \{(t, x) \in I\!R \times E; \; t \cdot d(x, E \setminus U) = 1\}$$

is closed. Therefore V is a Polish space and $V \ni (t, x) \to x \in U$ is a homeomorphism so U is a Polish space.

Let now Π be the family of subset E of X so that E and its complement are K-analytic. We have just shown that Π contains all open sets and by a) Π contains X. By Theorem II:1 b) Π is a σ-algebra and therefore Π contains all Borel sets.

Corollary II:1. Every analytic set is K-analytic.

Proof. Let P be a Polish space. By Theorem II:2a), there is a G_δ set $\bigcap_{j \in I\!N} O_j$ contained in a compact metric space so that

$f(\cap O_j)=P$ for a continuous f. In a compact metric space, every open set is a K_σ-set. Therefore P is K-analytic and so is any continuous image of P.

Theorem II:3. Every K-analytic set in U is universally capacitable. (Remember, U is assumed to be F_σ). For the proof, we need two lemmas.

Lemma II:1. Every $K_{\sigma\delta}$ is universally capacitable.

Proof. Assume that $A=\cap_n A_n$ where $A_n=\cup_p K_{n,p}$, $K_{n,p}$ compact and increasing in p. Let c be a given capacity and λ a given number $< c(A)$. Since $A \subset A_1$ there is a p_1 so that $c(A \cap K_{1,p_1})>\lambda$. Put $a_1=A \cap K_{1,p_1}$ and define $(a_n)_{n=1}^\infty$ inductively: If a_{n-1} is chosen, take p_n so big that $c(a_n)>\lambda$ where $a_n=a_{n-1} \cap K_{n,p_n}$.

Since

$$a_n \subset K_{1,p_1} \cap \ldots \cap K_{n,p_n}$$

we have that

$$c(K_{1,p_1} \cap \ldots \cap K_{n,p_n})>\lambda.$$

The set $K= \cap_{n=1}^\infty K_{n,p_n}$ is compact and contained in A since $K_{n,p_n} \subset A_n$.

Furthermore, since $c(K) \geq \lambda$ by Axiom iii), the lemma is proved.

Lemma II:2. If $f: U_0 \to U$ is a continuous function between two Hausdorff spaces and if c is a capacity on U then $c \circ f$ is a capacity on U_0.

Proof of Theorem II:3. Let A be a K-analytic set in a compact Hausdorff space U. By Proposition II:1, there is a compact space B_0 containing a $K_{\sigma\delta}$-set B and a continuous function f: $B_0 \to U$ so that A=f(B). If c is a given capacity on U, then cof is a capacity on B_0 by Lemma II:2. Since B is a $K_{\sigma\delta}$-set it follows from Lemma II:1 that c(f(B))=sup{c(f(K)): K compact subset of B}, which completes the proof since f is continuous on B_0.

Capacitability concerns inner regularity; approximation from the inside with compact sets.

Outer regularity; approximation from the outside with open sets, is the topic in the next section. In contrast to the case of measures, there are F_σ-sets that are not outer regular.

Notes and references

Theorem II:3 is due to Choquet.

Choquet, G., Theory of capacities. Ann. Inst. Fourier 5 (1953-54).

Choquet, G., Lectures on analysis. W.A. Benjamin. Inc. 1969.

See also chapter two in: **Federer, H.,** Geometric measure theory. Springer-Verlag, Berlin-Heidelberg-New York. 1969.

And the appendix in: **Treves, F.,** Topological vector spaces, distributions and kernels. Academic Press Inc. 1967.

There are analytic sets with complements that are not universally capacitable, see: **Dellacherie, D.,** Ensembles analytiques, capacités, mesures de Hausdorff. Springer LNM. 295, 1972, pg 28.

IIIa Outer Regularity

In this section, we assume S to be a compact and metric space.

Definition. Let c be a capacity on S. We say that c is <u>outer regular</u> if $c^*(E) = \inf\{c(O);\ E \subset O,\ O\ \text{open}\}$ is a capacity.

To verify that a given capacity is outer regular, it is enough to check property ii). Observe also that if c is outer regular, then it follows from Theorem II:3 that $c = c^*$ on all K-analytic sets, since they agree on all compacts.

It is clear that every positive measure defines an outer regular capacity. The following example shows that there exists a compact denumerable set M of probability-measures such that

i)   there is a F_σ-set F with $c(F) < c^*(F)$

ii) $c(E) = 0 \iff c^*(E) = 0$,   where $c(E) = \sup_{\mu \in M} \mu^*(E)$.

Example III:1. Let δ_0 and δ_n, $n \in \mathbb{N}$, be Dirac measures with mass at zero and $\frac{1}{n}$, respectively. Let m be the Lebesgue measure on the unit interval [0,1] and denote by M the denumerable compact set of measures $\{\delta_i \otimes m\}_{i=0}^{\infty}$ and let E be the F_σ-set

$$\{(0,y) \in R^2;\ \tfrac{1}{2} \le y \le 1\} \cup_{i=1}^{\infty} \{(\tfrac{1}{i},y);\ 0 \le y \le \tfrac{1}{2}\}.$$

If we, as usual, define $c(E) = \sup_{\mu \in M} \mu(E)$ we have $c(E) = \frac{1}{2}$ but $c^*(E) = 1$.

Definition. Let c be a capacity. Define $\bar{c}$ by

$$\bar{c}(E) = \inf\{c(F)'; \ E \subset F, \ F \ a \ G_\delta\text{-set}\}.$$

It is clear that $c(E) \leq \bar{c}(E) \leq c^*(E)$ and we now use Example III:1 to construct a capacity (of the type considered in Theorem I:1) such that $c \neq \bar{c}$.

Example III:2. Let M be the compact set of probability measures, $M = \{\delta_x \otimes m\}_{x \in [0,1]}$ where δ_x is the Dirac measure with mass at x and where m is the Lebesgue measure on $[0,1]$. We define the capacity c by $c(E) = \sup_{\mu \in M} \mu^*(E)$.

We first define E_n, $n \in \mathbb{N}$, by induction. Let E_1 be the set in Example III:1. To construct E_n, $n>1$, divide $[0,1]$ in n equal intervals and do the construction of Example III:1 in each interval so that the set so obtained does not intersect E_{n-1}. Put $E = \bigcup_{n=1}^{\infty} E_n$.

Then $c(E) = \frac{1}{2}$ but if F is any G_δ-set containing E we claim that $c(F) = 1$. For if $\bigcap_{s=1}^{\infty} 0_s = F$, $E \subset 0_s$, $s \in \mathbb{N}$ are open sets then the 1-dimensional sets $S_s = \{x \in [0,1], \{x\} \times [0,1] \subset 0_s\}$, $s \in \mathbb{N}$ are open and dense in $[0,1]$ by the construction of E. Hence, $\bigcap_{s=1}^{\infty} S_s$ is dense and in particular non-empty which means that $c(\bigcap_{s=1}^{\infty} 0_s) = c(F) = 1$.

Example III:3. Let c be the capacity defined in Example III:2. Then there exists a G_δ-set A contained in $[0,1] \times [0,1]$ such that

i) $c(A) = 0$

ii) $c(O) = 1$ for every open set O containing A.

This follows directly from the existence of a G_δ-set A contained in $[0,1] \times [0,1]$ such that

1) A is the graph of a lower semicontinuous function.

2) If K is a compact set in $[0,1] \times [0,1]$ with

$\text{proj}_1 K = [0,1]$ then $A \cap K \neq \emptyset$.

Definition. A set function c is called strongly subadditive if

$$c(K_1 \cup K_2) + c(K_1 \cap K_2) \leq c(K_1) + c(K_2)$$

for all compacts K_1, K_2.

Theorem III:1. Every strongly subadditive capacity on S is outer regular.

Lemma III:1. Assume that c is strongly subadditive and that $U_i \supset V_i$, $1 \leq i \leq h$ are open subset of V. Then

$$c\left(\bigcup_{i=1}^{n} U_i\right) - \sum_{i=1}^{n} c(U_i) \leq c\left(\bigcup_{i=1}^{n} V_i\right) - \sum_{i=1}^{n} c(V_i).$$

Proof of the lemma. The proof is by induction. Assume that $U_i \supset V_i$, $1 \leq j \leq n$, are open sets. We want to prove that

$$c\left(\bigcup_{j=1}^{n} U_j\right) + \sum_{j=1}^{n} c(V_i) \leq c\left(\bigcup_{j=1}^{n} V_i\right) + \sum_{j=1}^{n} c(U_i).$$

This is true if $n = 1$. If $n = 2$ we put $U = U_1$ and $V = V_1 \cup U_2$. Then $c(U_1 \cup U_2) + c(V_1) \leq c(U \cup V) + c(U \cap V) \leq c(U) +$ $+ c(V) = c(V_1 \cup U_2) + c(U_1)$.

On the other hand, if we put $U = U_2$ and $V = V_1 \cup V_2$ we get $c(U_2 \cup V_1) + c(V_2) \leq c(U \cup V) + c(U \cap V) \leq c(U) + c(V) = c(U_2) + c(V_1 \cup V_2)$.

Adding the inequalities gives

$$c(U_1 \cup U_2) + c(V_1) + c(V_2) \leq c(V_1 \cup V_2) + c(U_1) + c(U_2)$$

which proves the lemma for $n = 2$.

Assume now that the formula has been proved for n. We then prove it for $n + 1$. Put

$$U^1 = \bigcup_{j=1}^{n} U_j, \quad U^2 = U_{n+1}$$

$$V^1 = \bigcup_{j=1}^{n} V_j, \quad V^2 = V_{n+1}.$$

The case $n = 2$ then gives

$$c(U^1 \cup U^2) + c(V^1) + c(V^2) \leq c(V^1 \cup V^2) + c(U^1) + c(U^2)$$

and the induction assumption

$$c(\bigcup_{j=1}^{n} U_j) + \sum_{j=1}^{n} c(V_j) \leq c(\bigcup_{j=1}^{n} V_j) + \sum_{j=1}^{n} c(U_j).$$

Hence

$$c(\bigcup_{j=1}^{n+1} U_j) - \sum_{j=1}^{n+1} c(U_j) - c(\bigcup_{j=1}^{n+1} V_j) + \sum_{j=1}^{n+1} c(V_j) -$$

$$- (\bigcup_{j=1}^{n} V_j) - c(V_{n+1}) + c(\bigcup_{j=1}^{n+1} V_j) + c(\bigcup_{j=1}^{n} U_j) + c(U_{n+1}) -$$

$$- \sum_{j=1}^{n+1} c(U_j) - c(\bigcup_{j=1}^{n+1} V_j) + \sum_{j=1}^{n+1} c(V_j) = c(\bigcup_{j=1}^{n} U_j) +$$

$$+ \sum_{j=1}^{n} c(V_j) - c(\bigcup_{j=1}^{n} V_j) - \sum_{j=1}^{n} c(U_j) \leq 0$$

by the induction assumption.

Proof of the theorem. It is enough to prove that ii) holds true for $c*$. So assume that $E_s \nearrow E$, $s \to +\infty$. Given $\varepsilon > 0$, for every i there is an open set U_i containing E_i such that $c(U_i) - c*(E_i) \leq \dfrac{\varepsilon}{2^i}$.

By Lemma III:1 we have

$$c(\bigcup_{i=1}^{n} U_i) + \sum_{i=1}^{n} c*(E_i) \leq c*(\bigcup_{i=1}^{n} E_i) + \sum_{i=1}^{n} c(U_i)$$

so

$$c(\bigcup_{i=1}^{n} U_i) - c*(\bigcup_{i=1}^{n} E_i) \leq \sum_{i=1}^{n} c(U_i) - c*(E_i) \leq \varepsilon.$$

Hence

$$0 \leq c(\bigcup_{i=1}^{\infty} U_i) - c*(E) \leq \lim_{n \to +\infty} (c(\bigcup_{i}^{n} U_i) - c*(E_n)) =$$

$$= \lim_{n \to +\infty} (c(\bigcup_{i=1}^{n} U_i) - c*(\bigcup_{i=1}^{n} E_i)) \leq \varepsilon.$$

Hence $c*(E) \leq \lim_{n \to +\infty} c*(E_n)$ which proves the theorem.

Let M be a set of positive measures on S with mass less or equal to 1. We finish this section by proving a theorem that gives sufficient condition on M so that the set function $\sup_{\mu \in M} \mu(E)$ is outer regular on its zerosets.

Definition. Let N be the set of positive, lower semicontinuous functions φ on S with the property that to every $\varepsilon > 0$ there is an open set A with $\sup_{\mu \in M} \mu(A) < \varepsilon$ and such that the restriction of φ to $S \setminus A$ is continuous.

Lemma III:2. Let Q be a downward directed family of positive functions such that $\overline{M} \ni \mu \to \int \varphi d\mu$ is continuous for every $\varphi \in Q$. Then $\inf\limits_{\varphi \in Q} \sup\limits_{\mu \in \overline{M}} \int \varphi d\mu = \sup\limits_{\mu \in \overline{M}} \inf\limits_{\varphi \in Q} \int \varphi d\mu$.

Proof. It is clear that $\sup\limits_{\mu \in \overline{M}} \inf\limits_{\varphi \in Q} \int \varphi d\mu \leq \inf\limits_{\varphi \in Q} \sup\limits_{\mu \in \overline{M}} \int \varphi d\mu$. Let $\alpha > \sup\limits_{\mu \in \overline{M}} \inf\limits_{\varphi \in Q} \int \varphi d\mu$. Given $\mu \in \overline{M}$ there is a $\varphi \in Q$ with $\int \varphi d\mu < \alpha$ so $A_\varphi = \{\mu \in \overline{M}; \int \varphi d\mu < \alpha\}$, $\varphi \in Q$ is then an open covering of $\overline{M}$ and since $\overline{M}$ is compact we can choose $(\varphi_i)_{i=1}^{T}$ so that $\bigcup\limits_{i=1}^{T} A_{\varphi_i} \supset \overline{M}$. But since Q is downward directed there is an $\varphi \in Q$ which is dominated by all φ_i, $1 \leq i \leq T$. Therefore $\int \varphi d\mu < \alpha$, $\forall \mu \in \overline{M}$ so $\inf\limits_{\varphi \in Q} \sup\limits_{\mu \in \overline{M}} \int \varphi d\mu \leq \alpha$ which proves the lemma.

Theorem III:2. Assume that φ is bounded, positive and lower semicontinuous. Then $\varphi \in N$ if and only if

$$\overline{M} \ni \mu \mapsto \int \varphi d\mu$$

is continuous.

Proof of Theorem III:2. $\Rightarrow$) Assume that $\varphi \in N$ and that $\mu_s \overset{\Rightarrow}{\to} \mu$. Given $\varepsilon > 0$ choose 0_ε as in the definition and extend φ to φ_ε, continuous on U. Then

$$\left| \int \varphi d\mu_s - \int \varphi d\mu \right| = \left| \int (\varphi - \varphi_\varepsilon) d\mu_s - \int (\varphi - \varphi_\varepsilon) d\mu + \int \varphi_\varepsilon d\mu_s - \right.$$

$$\left. - \int \varphi_\varepsilon d\mu \right| \leq 2\sup|\varphi|\varepsilon + 2\sup\varphi\varepsilon + \left| \int \varphi_\varepsilon d\mu_s - \varphi_\varepsilon d\mu \right|$$

so $\overline{\lim_{s\to 0}}\,|\int\varphi d\mu_s - \int\varphi d\mu| = 0$. (Note that $\mu(0_\varepsilon) \leq \varepsilon$ by Lemma I:1).

$\Leftarrow$) Assume that $\overline{M} \ni \mu \mapsto \int\varphi d\mu$ is continuous and let N_1 be the positive continuous functions dominated by φ. By Lemma III:2 we have $0 = \sup_{\mu\in M} \inf_{\Psi\in N_1} \int(\varphi-\Psi)d\mu = \inf_{\Psi\in N_1} \sup_{\mu\in\overline{M}} \int(\varphi-\Psi)d\mu$. We can choose $\Psi_i \in N_1$, $i \in \mathbb{N}$, to be an increasing sequence of functions with $\sup_{\mu\in M} \int\varphi - \Psi_j d\mu < \dfrac{1}{2^{2j+1}}$, $j=1,2,\ldots$, and with limit φ.

Then $\varphi = \Psi_1 + \sum_{j=1}^{\infty} \Psi_{j+1} - \Psi_j$ and if we put $0_K = \{x: \sum_{j=K}^{\infty} \Psi_{j+1} - \Psi_j > \dfrac{1}{2^K}\}$

then $\mu(0_K) \leq \int 2^K (\sum_{j=K}^{\infty} \Psi_{j+1} - \Psi_j) d\mu \leq \dfrac{2^K}{2^{2K+2}} < \dfrac{1}{2^K}$. Let $E = \bigcup_{K=m}^{\infty} 0_K$.

Then $c(E) < \dfrac{1}{2^{m-1}}$ and on CE we have $\sum_{j=K}^{\infty} \Psi_{j+1} - \Psi_i \leq \dfrac{1}{2^K}$, $K \geq m$,

which shows that $\sum_{j=1}^{\infty} \Psi_{j+1} - \Psi_j$ converges uniformly on CE; φ is continuous on CE which completes the proof.

Theorem III:3. Let M and N be defined as above. Assume that N contains a convex cone R of functions with the following properties.

i) $1 \in R$.

ii) If $(\varphi_j)_{j=1}^{\infty}$ is a uniformly bounded and monotone sequence in R, then $\varphi_0 = (\lim_{j\to+\infty} \varphi_j)_* \in R$ and $\lim_{j\to+\infty}\int\varphi_j d\mu = \int\varphi_0 d\mu$, $\forall\mu \in M$ where φ_0 is the largest lower semicontinuous minorant of $\lim_{j\to+\infty} \varphi_j$.

iii) If $\varphi,\Psi \in R$ then $\inf(\varphi,\Psi) \in R$.

iv) If $(A_j)_{j=1}^{\infty}$ is a decreasing sequence of open sets with

$$\lim_{j\to+\infty} \sup_{\mu\in M} \mu(A_j) = 0 \quad \text{then}$$

$$\lim_{j \to +\infty} \inf\{\sup_{\mu \in M} \int \varphi d\mu; \ \varphi \in R, \ \varphi \geq 1 \ \text{on} \ A_j\} = 0.$$

v)   If K is a compact subset of S with $\sup_{\mu \in M} \mu(K) = 0$ then

there is a sequence $(A_j)_{j=1}^{\infty}$ of open sets containing K

such that

$$\lim_{j \to +\infty} \sup_{\mu \in M} \mu(A_j) = 0.$$

Then $G(E) = \inf\{\sup_{\mu \in M} \int \varphi d\mu; \ \varphi \in R, \ \varphi \geq 1 \ \text{on} \ E\}$ is an outer regular capacity.

Corollary III:1. Assume that M, N and R are as in Theorem
III:3. Then, to every Borel set E in S with $\sup_{\mu \in M} \mu(E) = 0$
there is a decreasing sequence $(A_j)_{j=1}^{\infty}$ of open sets containing
E with

$$\lim_{j \to +\infty} \sup_{\mu \in M} \mu(A_j) = 0.$$

Proof. Since all functions in N and hence in R are lower
semicontinuous, the set function

$$G(E) = \inf\{\sup_{\mu \in M} \int \varphi d\mu; \ \varphi \in R, \ \varphi \geq 1 \ \text{on} \ E\}$$

is "outer" in the sense that

$$G(E) = \inf\{G(A); \ E \subset A \ \text{open}\}.$$

This proves that G satisfies axiom iii) and also that the
corollary follows from the theorem.

Let now $(\varphi_j)_{j=1}^{\infty}$ be a decreasencee of functions
in R, $\varphi_j \leq 1$ and let φ_0 be the largest lower semicontinuous

minorant of $\lim\limits_{j\to+\infty} \varphi_j$. We claim that to every $\varepsilon > 0$ there is a $\varphi_\varepsilon \in R$ such that $\varphi_\varepsilon = 1$ on $\{\lim \varphi_j > \varphi_0\}$ and such that $\sup\limits_{\mu\in M} \int \varphi_\varepsilon d\mu < \varepsilon$. Let $\varepsilon > 0$ be given. Since all the functions φ_0, $(\varphi_j)_{j=1}^{\infty}$ belongs to N, there is a decreasing sequence of open sets $(A_j)_{j=1}^{\infty}$ with $\lim\limits_{j\to+\infty} \sup\limits_{\mu\in M} \mu(A_j) = 0$ and such that all the functions are continuous on CA_j, $j \in \mathbb{N}$. By iv), there is a $\varphi_\varepsilon \in R$ such that $\sup\limits_{\mu\in M} \int \varphi_\varepsilon d\mu < \varepsilon/3$, $\varphi_\varepsilon \geq 1$ on A_{j_ε} for some A_{j_ε}. Then $\{x \in CA_{j_\varepsilon} ; \varphi_j \geq \varphi_0 + \frac{1}{v}\} = K_j^v$ is a decreasing sequence of compact sets and $\sup\limits_{\mu\in M} \mu(\bigcap\limits_{j=1}^{\infty} K_j^v) = 0$ by ii). Hence by v) and iv) there is a sequence $(\Psi^v)_{v=1}^{\infty}$ of functions in R with $\Psi^v \geq 1$ on $\bigcap\limits_{j=1}^{\infty} K_j^v$ and such that

$$\sup_{\mu\in M} \int \Psi^v d\mu < \frac{\varepsilon}{6\cdot 2^v} \ .$$

Then $\tau = \inf(\sum\limits_{v=1}^{\infty} \Psi^v, 1) \in R$ by (ii) for increasing sequences, and $\tau \geq 1$ on $\{\varphi_0 < \lim \varphi_j\} \cap CA_{j_\varepsilon}$. Furthermore $\sup\limits_{\mu\in M} \int \tau d\mu < \frac{\varepsilon}{3}$ so $\varphi_\varepsilon + \tau \geq 1$ on $\{\varphi_0 < \lim \varphi_i\}$ and $\sup\limits_{\mu\in M} \int \varphi_\varepsilon + \tau < \varepsilon$ which proves the claim.

Let now $(E_j)_{j=1}^{\infty}$ be an increasing sequence of subsets of S with $E = \bigcup\limits_{j=1}^{\infty} E_j$. We want to prove that $\lim\limits_{j\to+\infty} G(E_j) = G(E)$. Choose $\varphi_K^j \in R$, $\varphi_K^j \geq 1$ on E_j so that $\sup\limits_{\mu\in M} \int \varphi_K^j d\mu \searrow G(E_j)$, $K\to+\infty$, $\forall j \in \mathbb{N}$ where we can assume that $\varphi_{K+1}^j \leq \varphi_K^j$, $j,K \in \mathbb{N}$. We denote by φ_0^j the largest lower semicontinuous minorant of $\lim\limits_{K\to+\infty} \varphi_K^j$.

We can assume that $\varphi_0^j \leq \varphi_0^{j+1}$, $j \in \mathbb{N}$ (for we can replace φ_K^j by $\min\{\varphi_m^l;\ l \geq j,\ l + m \leq j + K\}$). Let $\varepsilon > 0$ be given and choose φ_ε^j as above so that $\varphi_\varepsilon^j = 1$ where $\lim\limits_{K \to +\infty} \varphi_K^j > \varphi_0^j$ and so that $\sup\limits_{\mu \in M} \int \varphi_\varepsilon^j d\mu < \dfrac{\varepsilon}{2^j}$.

Then $\Psi_j = \varphi_0^j + \inf(\sum\limits_{s=1}^{\infty} \varphi_\varepsilon^s, 1) \in R$, $\Psi_j \leq \Psi_{j+1}$ and $\Psi_j \geq 1$ on E_j. Hence $G(E) \leq \sup\limits_{\mu \in M} \int \lim\limits_{j \to +\infty} \Psi_j d\mu = \sup\limits_{\mu \in M} \lim\limits_{j \to +\infty} \int \Psi_j d\mu =$

$= \sup\limits_{\mu \in M} \lim\limits_{j \to +\infty} \int (\varphi_0^j + \inf(\sum\limits_{s=1}^{\infty} \varphi_\varepsilon^s, 1)) d\mu \leq \lim\limits_{j \to +\infty} \sup\limits_{\mu \in M} \int \varphi_0^j d\mu + \varepsilon$. But since all functions $(\varphi_0^j)_{j=1}^{\infty}$ and $(\varphi_K^j)_{j,K=1}^{\infty}$ are lower semi-continuous we have by ii)

$$\sup\limits_{\mu \in M} \int \varphi_0^j d\mu = \sup\limits_{\mu \in M} \inf\limits_K \int \varphi_K^j d\mu \leq \inf\limits_K \sup\limits_{\mu \in M} \int \varphi_K^j d\mu \leq G(E_j).$$

Hence $G(E) \leq \lim\limits_{j \to +\infty} G(E_j) + \varepsilon$ which proves the theorem.

Remark. If R and M satisfies i)-v) then M can be replaced by its weak*-closure.

Notes and references

Example III:1 is due to **B. Fuglede,** Capacity as a sublinear functional generalizing an integral. Der Kongelige Danske Videnskabernes Selskab. Matematisk-fysiske Meddelelser. 38.7 (1971).

The existence of a G_δ-set A with properties 1) and 2) in Example III:3 was proved by **Roy O. Davies,** A non-Prokhorov space, Bull. London Math. Soc. 3 (1971), 341-342. The use of A in this context was observed by **C. Dellacherie,** Ensembles analytiques, capacités, mesures de Hausdorff. Springer LNM. 295, 1972. pg. 106 Ex. 4.

Theorem III:1 is a variant of a theorem due to Choquet. See the references in Section II.

Representation of strongly subadditive capacities by measures has been studied by **Bernd Anger,** Representation of capacities. Math. Ann. 229 (1977), 245-258.

IIIb Outer Regularity (Cont.)

In this section, we continue our study of outer regularity but in a more special situation. Many problems in complex function theory are related to outer regular capacities - in particular outer regularity of zero sets. We therefore proceed as follows.

Assumptions. Let in what follows F be a convex cone of positive and lower semicontinuous functions (l.s.c.) defined on a compact and metric space U.

$h_g = \inf\{\varphi \in F;\ g \le \varphi\}.$

$H_g = \sup\{\Theta;\ \Theta \text{ continuous},\ \Theta \le g\}$

where we assume that $H_{h_g} \in F$ for every bounded and positive function g and that h_g is continuous if g is.

Let δ be a given probability measure on U such that $\int H_{h_g} d\delta = \int h_g d\delta$ for all bounded positive functions g; we also assume that $\int \varphi d\delta < +\infty,\ \forall \varphi \in F.$

Furthermore, we assume that if $\{\varphi_i\}_{i=1}^{\infty}$ is an increasing sequence of functions in F with $\lim_{i \to +\infty} \int \varphi_i d\delta < +\infty$ then $\lim_{i \to +\infty} \varphi_i \in F.$

Observe that $H_{h_g} = h_g$ for all l.s.c. g and that

$H_\varphi = h_\varphi = \varphi$ for all $\varphi \in F$. Note also that $\varphi_1, \varphi_2 \in F$ implies $\inf(\varphi_1, \varphi_2) \in F.$

For if g is l.s.c., choose $g_j \nearrow g$, g_j continuous. Then $h_{g_j} = H_{h_{g_j}} \leq H_{h_g} \leq h_g$.

Now $h_{g_j} \in F$; $h_{g_j} \nearrow$ so $\lim h_{g_j} \in F$ and since $\lim h_{g_j} \geq g$ we get that $h_g = \lim h_{g_j} \in F$ and that $h_g = H_{h_g}$.

The "fine" problem is now to decide if $E \mapsto h_{\chi_E}(z)$ is a capacity for every fixed $z \in U$ (χ_E is the characteristic function for E).

The "coarse" problem is to decide if $E \mapsto \int h_{\chi_E}(z)d\delta(z)$ is a capacity.

Assuming all this about F, we define a class of positive measures M, $M = \{\mu \geq 0; \int \varphi d\mu \leq \int \varphi d\delta, \forall \varphi \in F\}$.

It is clear that M is convex and since every function in F is l.s.c., M is compact by Lemma I:1. We now define c, $c(E) = \sup_{\mu \in M} \mu(E)$ which is a capacity by Theorem I:1, and the connection with outer regularity is that c outer regular if and only if $E \mapsto \int h_{\chi_E} d\delta$ is a capacity (cf. Proposition III:1 below).

We now turn to the study of the following statements.

1) Every bounded function in F is a member of N.

2) c is outer regular.

3) $c(E) = \int h_{\chi_E} d\delta$ for every Borel set E.

4) If E is a Borel set with $c(E) = 0$ then $c*(E) = 0$.

5) $c(\{h_g > H_{h_g}\}) = 0$ for every positive and bounded function g.

Lemma III:3. Define for bounded functions g:

$$c(g) = \sup_{\mu \in M} \int g \, d\mu \quad \text{and} \quad L(g) = \int h_g \, d\delta.$$

Then

1) $c(g) \leq L(g)$.

2) Equality holds in 1) if g is upper or lower semicontinuous.

3) $L(g) = \inf\{L(\varphi); \ g \leq \varphi \in \text{l.s.c.}\} = \inf\{L(\varphi); \ g \leq \varphi \in F\}$.

Proof. 1) Assume $g \geq 0$. Since $\int h_g \, d\delta = \int H_{h_g} \, d\delta$, there is, by Choquet's lemma, $\varphi_i \geq h_g$, $\varphi_i \in F$, $i \in \mathbb{N}$, a decreasing sequence of functions such that $\int \varphi_i \, d\delta \rightarrow \int h_g \, d\delta$, $i \rightarrow \infty$.

Thus, if $\mu \in M$; $\int g \, d\mu \leq \int h_g \, d\mu \leq \int \varphi_i \, d\mu \leq \int \varphi_i \, d\delta$ which gives

$$c(g) = \sup_{\mu \in M} \int g \, d\mu \leq \int h_g \, d\delta = L(g).$$

2) It is clear that the functional L has the following properties:

i) $L(\alpha g) = \alpha L(g)$, $\alpha \geq 0$.

ii) $L(g_1 + g_2) \leq L(g_1) + L(g_2)$.

iii) If $0 \leq g_1 \leq g_2$ then $L(g_1) \leq L(g_2)$.

From i), ii) and the Hahn-Banach theorem it follows that to every continuous function g there is a measure s such that

$$\int g \, ds = L(g)$$

$$\int \varphi \, ds \leq L(\varphi), \quad \forall \ \text{continuous} \ \varphi.$$

Thus if $s = s^+ - s^-$ is the decomposition of s in positive and negative parts, it follows from iii) that $\int \varphi \, ds^+ \leq L(\varphi)$, $\forall$ continuous φ.

Assume that $\Psi \in F$: choose $\{\Psi_i\}_{i=1}^{\infty}$ and increasing sequence of continuous functions with limit $= \Psi$. Then $\int \Psi ds^+ = \lim \int \Psi_i ds^+ \leq \lim L(\Psi_i) \leq L(\Psi)$ we have proved that $s^+ \in M$ so $c(g) = L(g)$, for all continuous g. If g is upper semicontinuous, choose $\{g_i\}_{i=1}^{\infty}$ to be a decreasing sequence of continuous function with limit $= g$. Then $L(g) \leq \lim_i L(g_i) = \lim_i c(g_i) \leq \lim_i \mu_i(g_i) \leq \lim \mu_i(g_k) \leq \mu(g_k)$ where we can assume that $\mu_i \searrow \mu$. This gives $L(g) \leq \mu(g) \leq c(g)$. If g is lower semicontinuous then $g_i \nearrow g$; $H_{h_{g_i}} = h_{g_i} \nearrow h_g = H_{h_g}$ so

$$L(g) = \int h_g d\delta = \lim_i \int h_{g_i} d\delta = L(g_i) = c(g_i) \leq c(g), \quad i \to +\infty.$$

Finally, in order to prove 3) use Choquet's lemma and choose $\varphi_i \in F$, $i \in \mathbb{N}$, $\varphi_i \geq g$ so that if Θ is l.s.c. with $\Theta \leq \inf_{i \in \mathbb{N}} \varphi_i$ then $\Theta \leq h_g$. In particular $H_{h_g} = H_{\inf_{i \in \mathbb{N}} \varphi_i}$ so since $H_{h_g} \leq h_g \leq h_{\inf_{i \in \mathbb{N}} \varphi_i} = \inf_{i \in \mathbb{N}} \varphi_i$ we get

$$\int H_{\inf_{i \in \mathbb{N}} \varphi} d\delta = \int \inf_{i \in \mathbb{N}} \varphi_i d\delta \quad \text{so} \quad h_g = \inf_{i \in \mathbb{N}} \varphi_i \quad \text{a.e. } (d\delta).$$

In other words: $\int h_g d\delta = \lim_{j \to +\infty} \int \inf_{1 < i < j} \varphi_i d\delta$ which completes the proof.

Proposition III:1. $2) \iff 3)$.

Proof. $2) \Rightarrow 3))$ By Lemma III:3, $c(E) = L(\chi_E)$ if E is open or closed. So if c is outer regular then $c(E) \leq L(\chi_E) \leq \inf_{\substack{E \subset 0 \\ 0 \text{ open}}} L(\chi_0) = \inf_{E \subset 0} c(0) = c(E)$.

On the other hand, 3) together with Lemma III:3 3) proves that 2) holds.

Corollary III:1. 3) $\Rightarrow$ 4).

Lemma III:4. We have that $c*(E)=0 \iff$ there is a $\varphi \in F$ with $\int \varphi d\delta < +\infty$ but $\varphi|_E = +\infty$

Proof. $\Leftarrow$) Put $0_\varepsilon = \{z \in U;\ \varepsilon\varphi(z)>1\}$, $\varepsilon>0$.

Then 0_ε is open, contains E and $c(0_\varepsilon) \leq \varepsilon \int \varphi d\delta \to 0$, $\varepsilon \to 0$.

$\Rightarrow$) Choose for every $s \in \mathbb{N}$, 0_s open with $C(0_s) < \dfrac{1}{2^s}$. Then $h_{0_s} = H_{h_{0_s}}$, which we denote by f_s, is in F.

Then $c(0_s) = \int f_s d\delta$ so $\varphi = \sum\limits_{s=1}^{\infty} f_s$ is the required function.

Proposition III:2. 4) + 5) $\Rightarrow$ 3).

It is enough to prove that $E \to \int h_{\chi_E} d\delta$ is a capacity since $c(K) = \int h_{\chi_K} d\delta$ for all compact sets K by Lemma III:3. Let $E_s \nearrow E$ be a given increasing sequence of sets and $\varepsilon>0$. By Lemma III:4 there is a $\Psi_s \in F$ such that $\{h_{\chi_{E_s}} > H_{\chi_{E_s}}\} \subset \{\Psi_s = +\infty\}$ where we can assume that $\int \Psi_s d\delta < \dfrac{1}{2^s}$. Put $F = \sum\limits_{s=1}^{\infty} \Psi_s$; we then have $H_{h_{\chi_{E_s}}} \leq h_{\chi_{E_s}} \leq H_{h_{\chi_{E_s}}} + \varepsilon F$ so $H_{h_{\chi_{E_s}}} \leq H_{h_{\chi_E}} \leq \lim\limits_{s \to +\infty} H_{h_{\chi_{E_s}}} + \varepsilon F$.

Hence $\lim \int H_{h_{\chi_{E_s}}} d\delta \leq \int H_{h_{\chi_E}} d\delta \leq \lim\limits_{s} \int H_{h_{\chi_{E_s}}} d\delta + \varepsilon \int F d\delta$. Thus $\lim\limits_{s \to \infty} \int h_{\chi_{E_s}} d\delta = \int h_{\chi_E} d\delta$ which means that $E \to \int h_{\chi_E} d\delta$ is a capacity and the proposition follows.

Proposition III:3. 1) + 5) $\Rightarrow$ 3).

Proof. Follows from Theorem III:3. In this situation, there is a simpler proof. It follows from the proof of Proposition III:2 that if $c*(\{h_g > H_{h_g}\}) = 0$ for all positive and bounded g then 3) holds true. Choose $\varphi_i \in F$ to be a decreasing sequence with $\lim_{i \in N} \varphi_i = H_{h_g}$ a.e. $(d\delta)$. Given $\varepsilon > 0$, choose 0_ε open with $c(0_\varepsilon) < \varepsilon$ such that φ_i, $i \in \mathbb{N}$ and H_{h_g} is continuous on $U \setminus 0_\varepsilon$. Then $\{h_g > H_{h_g}\} \subset 0_\varepsilon \cup \bigcup_{s=1}^{\infty} N_s^\varepsilon$ where

$$N_s^\varepsilon = \{x \notin 0_\varepsilon : h_g(x) \geq H_{h_g}(x) + \tfrac{1}{s}\}.$$

Each N_s^ε is closed so $c*(N_s^\varepsilon) = 0$ which means that $\{h_g > H_{h_g}\}$ can be covered by open sets of arbitrary small capacity which proves the proposition.

Theorem III:4. Assume that φ is a bounded function in F. Then $\varphi \in N$ if and only if there are two sequences of continuous functions in F, $(u^p)_{p=1}^{\infty}$ and $(u_p)_{p=1}^{\infty}$ such that

i) $u^p \nearrow \varphi$, $p \to +\infty$.

ii) $F \ni (u^p + \sum_{t=p}^{\infty} u_t) \searrow \varphi$, $p \to +\infty$ outside a set E with $c*(E) = 0$.

Proof. Assume first that $\varphi \in F \cap N \cap L^{\infty}$. As in the proof of Theorem III:2 we can choose an increasing sequence $(\Psi_j)_{j=1}^{\infty}$ of continuous functions in F with $\lim_{j \to +\infty} \Psi_j = \varphi$ and $\sup_{\mu \in M} \int (\varphi - \Psi_j) < \tfrac{1}{2^j}$. Lemma III:3 2) gives $\int H_{h_{(\varphi - \Psi_j)}} d\delta < \tfrac{1}{2^j}$.

Define $u^p = \Psi_p$ and

$$u_p = H_{h_{(\Psi_{p+1} - \Psi_p)}}.$$

Then all the functions are continuous and in F.

Since $\Psi_{j+1} - \Psi_j \leq H_{h_{\Psi_{j+1} - \Psi_j}}$ we get $u^{p+1} + \sum_{t=p+1}^{\infty} u_t - u^p -$

$- \sum_{t=p}^{\infty} u_t = u^{p+1} - u^p - u_p = \Psi_{p+1} - \Psi_p - H_{h_{\Psi_{p+1} - \Psi_p}} \leq 0$ so that

$u^p + \sum_{t=p}^{\infty} u_t$ is a decreasing sequence. Furthermore

$$u^p \leq \varphi \leq u^p + \sum_{t=p}^{\infty} u_t \quad \text{so}$$

$\{\lim u^p + \sum_{t=p}^{\infty} u_t > \varphi\} \subset \{\sum_{t=1}^{\infty} u_p = +\infty\}$ which completes the proof in

this direction since $\int \sum_{j=1}^{\infty} u_j d\delta < +\infty$.

On the other hand, if $(u^p)_{p=1}^{\infty}$ and $(u_p)_{p=1}^{\infty}$ have the

properties above, we wish to prove that $\varphi \in N$. Given $\varepsilon > 0$,

choose 0_ε with $c(0_\varepsilon) < \varepsilon$ and such that $\varphi = \lim_{p \to +\infty} (u^p + \sum_{t=p}^{\infty} u_t)$

outside 0_ε. If $\mu_s \in M$, $\mu_s \rightharpoonup \mu$ then $\int \varphi d\mu \leq \varliminf_{s \to +\infty} \int_U \varphi d\mu_s \leq$

$\leq \varlimsup_{s \to +\infty} \int_U \varphi d\mu_s \leq \varepsilon \sup_{x \in U} \varphi(x) + \varlimsup_{s \to +\infty} \int_{C0_\varepsilon} \varphi d\mu_s \leq \varepsilon \sup_{x \in U} \varphi(x) +$

$\varlimsup_{s \to +\infty} \int_{C0_\varepsilon} (u^p + \sum_{t=p}^{\infty} u_t) d\mu_s \leq \varepsilon \sup_{x \in U} \varphi(x) + \int_{C0_\varepsilon} u^p d\mu +$

$$+ \varlimsup_{s \to +\infty} \int_{CO_\varepsilon} \sum_{t=p}^{\infty} u_t d\mu_s \leq \varepsilon \sup_{x \in U} \varphi(x) + \int_U u^p d\mu + \sum_{t=p}^{\infty} \int_U u^t d\delta.$$

Since $\sum_{t=p}^{\infty} u_t \in F$, $\sum_{t=p}^{\infty} \int_U u_t d\delta < +\infty$ so letting p tend to $+\infty$

the right hand side tends to $\varepsilon \sup_{x \in U} \varphi(x) + \int_U \varphi d\mu$ which proves

the theorem, using Theorem III:2.

Notes and references

A proof of "Choquet´s lemma" can be found in **Doob, J.L.,**
Classical potential theory and its probabilistic counterpart,
Springer-Verlag (1984).

IV Subharmonic Functions in $\mathbb{R}^n$

Let B be the unit ball in $\mathbb{R}^n$, let F be the restriction to B of all positive superharmonic functions on RB, where R is a fixed number >1. If we take δ to be the Lebesgue measure on B, it is well known that U=B and δ satisfies all the assumptions made in Section III; the "coarse" problem has a positive solution. But much more can be said: the fine problem has a positive solution.

Fix $x \in U$ and define $d_x(K)=\inf\{\varphi(x) \in F; \; \varphi \geq 1 \text{ on } K\}$ for K compact. Consider the class M_x of positive measures on $\overline{B}$, $m_x=\{\mu \geq 0; \int \varphi d\mu \leq \varphi(x), \; \forall \varphi \in F\}$. Then, by Lemma I:1 and Theorem I:1, M_x gives rise to a capacity:

$$c_x(E) = \sup_{\mu \in M_x} \mu(E)$$

and by a proof, similar to that of Lemma III:3 we have that $c_x(E) \leq d_x(E)$ with equality if E is compact or open.

Furthermore, it is a consequence of the maximum principle for harmonic functions that $d_x(K_1 \cup K_2) + d_x(K_1 \cap K_2) \leq d_x(K_1) + d_x(K_2)$ so by Theorem III:1, d_x extends to an outer regular capacity and therefore $c_x=d_x$.

Moreover, the following stronger version of Theorem III:4 holds true in this case. If f is a bounded and positive superharmonic function on RB then

$$f = \sum_{j=1}^{\infty} f_j$$

where $(f_j)_{j=1}^{\infty}$ is a sequence of positive and continuous super-harmonic functions on B.

Notes and references

Brelot, M., Eléments de la théorie classique du potentiel. Centre documentation Universitair le Sorbonne, 1965.

Choquet, G., Theory of capacities. Ann. Inst. Fourier 5 (1953-54).

Choquet, G., Lectures on analysis. W.A. Benjamin. New York and Amsterdam 1969.

Landkof, N.S., Foundation of modern potential theory, Springer-Verlag, 1972.

V Plurisubharmonic Functions in $\mathbb{C}^n$ – The Monge-Ampère Capacity

Let B be the unit ball in $\mathbb{C}^n$ and let F be the restriction to B of all positive plurisuperharmonic functions on RB, where R is a fixed number >1. We take δ to be the Lebesgue measure on B and form c as in Section III:b. It is then true that F and δ meet all the requirements in Section III:b and we are going to see that F has property 1) and c has property 5); thus 2) and 3) hold true by Propositions III:3 and III:1.

The differential operators ∂ and $\overline{\partial}$ are defined by

$$\partial = \sum_{j=1}^{n} \frac{\partial}{\partial z_j} \, dz_j \quad \text{and} \quad \overline{\partial} = \sum_{j=1}^{n} \frac{\partial}{\partial \overline{z}_j} \, d\overline{\partial}_j \quad \text{so that} \quad d=\partial+\overline{\partial} \quad \text{and}$$

$d^c = i(\partial-\overline{\partial})$.

a. <u>Definition of the Monge-Ampère operator</u>

Let U be an open and bounded subset of $\mathbb{C}^n$. If $v^1,\ldots,v^n \in C^2(U)$, we define $MA(v^1,\ldots,v^n)$ to be the symmetric and n-linear operator

$$MA(v^1,\ldots,v^n) = dd^c v^1 \wedge \ldots \wedge dd^c v^n.$$

If moreover $v^1,\ldots,v^n \in PSH(U)$ then $dd^c v^1 \wedge \ldots dd^c v^n$ is a positive measure.

Theorem V:1. If $PSH \cap C^2(U) \ni v_j^i \searrow v^i \in L^\infty(U)$, $j \to +\infty$, $1 \leq i \leq n$ then $MA(v_j^1, \ldots, v_j^n)$ is a weakly convergent sequence of positive measures. Moreover, the limit is independent of the particular choice of the decreasing sequences v_j^i, $1 \leq i \leq n$; $j \in \mathbb{N}$.

Proof. The statement is purely local, so we can assume that $v_1^i = v_j^i$, $1 \leq i \leq n$, $j \in \mathbb{N}$ outside a fix compact subset K of $U = B$; the unit ball in $\mathbb{C}^n$.

Given $\Theta \in C_0^\infty(B)$, we want to prove that

$$\lim_{j \to +\infty} \int \Theta dd^c v_j^1 \wedge \ldots \wedge dd^c v_j^n \quad \text{exist. Take} \quad \Theta_1, \Theta_2 \in PSH \cap C^\infty(B), \quad \Theta = \Theta_1 - \Theta_2$$

and choose $0 \leq \eta \in C_0^\infty(B)$, $\eta = 1$ near the support of Θ, union K.

Then by Stokes formula,

$$\int \Theta dd^c v_j^1 \wedge \ldots \wedge dd^c v_j^n = \int v_j^1 \wedge dd^c \Theta \, dd^c v_j^2 \wedge \ldots dd^c v_j^n =$$

$$= \int \eta v_j^1 dd^c (\Theta_1 - \Theta_2) dd^c v_j^2 \wedge \ldots dd^c v_j^n.$$

We consider $\int \eta v_j^1 dd^c \Theta_1 \wedge dd^c v_j^2 \wedge \ldots dd^c v_j^n$ and observe that

$$\int \eta v_{j+1}^1 \wedge dd^c \Theta_1 \wedge \ldots dd^c v_{j+1}^n \leq \int \eta v_j^1 dd^c \Theta_1 \wedge \ldots dd^c v_{j+1}^n \leq$$

$$\leq \int v_{j+1}^2 dd^c \Theta_1 \wedge dd^c (\eta v_1^j) \wedge \ldots \wedge dd^c v_{j+1}^n =$$

$$= \int (v_{j+1}^2 - v_j^2) dd^c \Theta_1 \wedge dd^c (\eta v_1^j) \wedge \ldots \wedge dd^c v_{j+1}^n +$$

$$+ \int v_j^2 \wedge dd^c\Theta_1 \wedge dd^c\eta v_1^j \wedge \ldots \ dd^c v_{j+1}^n, \quad \text{since} \quad \eta=1 \quad \text{near}$$

$$\text{supp } v_{j+1}^2 - v_j^2$$

we get $\quad \int (v_{j+1}^2 - v_j^2) dd^c\Theta_1 \wedge dd^c v_1^j \wedge \ldots \wedge dd^c v_{j+1}^n +$

$$+ \int \eta v_1^j dd^c\Theta_1 \wedge dd^c v_j^2 \wedge dd^c v_{j+1}^3 \wedge \ldots \wedge dd^c v_{j+1}^n \leq$$

$$\leq \int \eta v_1^j \wedge dd^c\Theta_1 \wedge dd^c v_j^2 \wedge dd^c v_{j+1}^3 \wedge \ldots \wedge dd^c v_{j+1}^n.$$

Repeating this, we get that

$$0 \leq \int \eta v_1^j dd^c\Theta_1 \wedge dd^c v_j^2 \wedge \ldots \wedge dd^c v_j^n$$

is decreasing in j which proves that the limit exist.

We now prove that the limit does not depend on the particular choice of the approximating sequences. We first note that it follows from the proof above that if $v_{j_p}^i$, $1 \leq i \leq n$, $j_p \in \mathbb{N}$ are subsequences of v_j^i, then $dd^c v_{j_p}^1 \wedge \ldots dd^c v_{j_p}^n$ tends weakly to the same limit as $dd^c v_j^1 \wedge \ldots \wedge dd^c v_j^n$. Since MA is n-linear, it is enough to show that $dd^c v_j^1 \wedge \ldots \wedge dd^c \tilde{v}_j^n$ tends to the same limit as $dd^c v_j^1 \wedge \ldots \wedge dd^c v_j^n$ if $\tilde{v}_j^n$ has the same properties as v_j^n. But if $\Theta \in C_0^\infty(B)$, we have that

$$\int \Theta dd^c v_j^1 \wedge \ldots \wedge dd^c v_j^{n-1} \wedge dd^c(v_k^n - \tilde{v}_k^n)$$

$$= \int (v_K^n - \tilde{v}_k^n) dd^c v_j^1 \wedge \ldots \wedge dd^c v_j^{n-1} \wedge dd^c\Theta$$

and since $v_k^n - \tilde{v}_k^n \to 0$, $k \to +\infty$ the right hand side tends to zero when $k \to +\infty$ for every fixed j which proves the theorem.

Definition. If $v^i \in PSH \cap L^\infty(U)$, $1 \leq i \leq n$ we define $MA(v^1, \ldots, v^n) = dd^c v^1 \wedge \ldots \wedge dd^c v^n$ to be the weak limit of $dd^c v_j^1 \wedge \ldots \wedge dd^c v_j^n$ where $PSH \cap C^2(U) \ni v_j^i \searrow v^i$, $j \to +\infty$, $1 \leq i \leq n$.

Definition. If E is a Borel subset of U, we put $d(E) = \sup\{\int_E MA(u, \ldots, u) : u \in PSH(U), 0 < u < 1\}$. It is easy to see $\underset{n \text{ times}}{}$ that the following proposition holds.

Proposition V:1.

i) $\quad E_1 \subset E_2 \Rightarrow d(E_1) \leq d(E_2)$

ii) $\quad d(\bigcup_{i=1}^{\infty} E_i) \leq \sum_{i=1}^{\infty} d(E_i)$

iii) If $E_j \nearrow E$: $j \to +\infty$ then $\lim d(E_j) = d(E)$.

(We will see later that d is a capacity in Choquet´s sense).

Theorem V:2. If $PSH \ni v_j^i \searrow v^i \in PSH \cap L_{loc}^\infty(U)$, $j \to +\infty$, $0 \leq i \leq n$ then

$$v_j^0 dd^c v_j^1 \wedge \ldots \wedge dd^c v_j^n \searrow v^0 dd^c v^1 \wedge \ldots dd^c v^n.$$

Proof. The statement is purely local, so we can assume that $v_1^0 = v_j^i$, $0 \leq i \leq n$; $j \in \mathbb{N}$ outside a fixed compact set K of $U = B$. Assume first that $v_j^i \in PSH \cap C^2(B)$. If $0 \leq \eta \in C_0^\infty(B)$, $\eta = 1$ near K, we find as in the proof of Theorem IV:1, that

$$\int \eta v_{j_0}^0 dd^c v_{j_1}^1 \wedge \ldots \wedge dd^c v_j^n$$

is decreasing in $(j_0, \ldots, j_n)$. Therefore

$$\lim_{\substack{\longrightarrow \\ j \to +\infty}} \int \eta v_j^0 dd^c v_j^1 \wedge \ldots dd^c v_j^n \geq \lim_{\substack{\longrightarrow \\ j \to +\infty}} \int \eta v_j^0 \, dd^c v_j^1 \wedge \ldots \wedge dd^c v^n \geq$$

$$\geq \ldots \geq \lim_{\substack{\longrightarrow \\ j \to +\infty}} \int \eta v_j^0 \, dd^c v^1 \wedge \ldots \wedge dd^c v^n = \int \eta v^0 dd^c v^1 \wedge \ldots \wedge dd^c v^n.$$

On the other hand, since ηv_j^0 is decreasing, we get

$$\overline{\lim_{j \to +\infty}} \int \eta v_j^0 dd^c v_j^1 \wedge \ldots \wedge dd^c v_j^n \leq \overline{\lim_{k \to +\infty}} \int \eta v_K^0 dd^c v^1 \wedge \ldots \wedge dd^c v^n \leq$$

$$\leq \int \eta v^0 dd^c v^1 \wedge \ldots \wedge dd^c v^n.$$

Therefore $\displaystyle \lim_{j \to +\infty} \int \eta v_j^0 dd^c v_j^1 \wedge \ldots \wedge dd^c v_j^n = \int \eta v^0 dd^c v^1 \wedge \ldots \wedge dd^c v^n$

so if μ is any accumulation point for

$$v_j^0 dd^c v_j^1 \wedge \ldots dd^c v_j^n, \quad j \in \mathbb{N}$$

μ has the same mass as $v^0 dd^c v^1 \wedge \ldots dd^c v^n$ and since Θu^0 is
upper semicontinuous if $0 \leq \Theta \in C_0^\infty(B)$

$$\int \Theta d\mu \leq \int \Theta v^0 dd^c v^1 \wedge \ldots dd^c v^n.$$

Thus $Z \leq u^0 dd^c v^1 \wedge \ldots dd^c v^n$ and they have the same mass so
they have to be equal.

Now, if v_j^i are in $PSH \cap L_{loc}^\infty(B)$ only, we consider the
regularized functions $v_{j,\varepsilon}^i \in PSH \cap C^\infty$ in a smaller ball.

Then, if $(\varepsilon_j)_{j=1}^\infty$ is any sequence decreasing to zero we get from above that

$$v^0_{i,\epsilon_i}\, dd^c v^1_{i,\epsilon_i} \wedge \ldots dd^c v^n_{i,\epsilon_i} \rightrightarrows v^0\, dd^c v^1 \wedge \ldots dd^c v^n.$$

On the other hand, given $\Theta \in C^\infty_0(B)$ we can to every j find ϵ_{p_j} so that

$$\left| \int \Theta v^0_{j,\epsilon_{p_j}}\, dd^c v^1_{j,\epsilon_{p_j}} \wedge \ldots \wedge dd^c v^n_{j,\epsilon_{p_j}} \right. -$$

$$\left. - \int \Theta v^0_j\, dd^c v^1_j \wedge \ldots \wedge dd^c v^n_j \right| < \frac{1}{j}$$

because the theorem is already proved for C^2-functions. Therefore $v^0_j dd^c v^1_j \wedge \ldots \wedge dd^c v^n_j \rightrightarrows v^0 dd^c v^n \wedge \ldots \wedge dd^c v^n$ which proves the theorem.

b.   <u>Quasicontinuity with respect to   d.</u>

Theorem V:3.   To every $\epsilon > 0$ and every $v \in PSH(U)$ there is an open set O_ϵ such that $d(O_\epsilon) < \epsilon$ and $v\big|_{U-O_\epsilon}$ is continuous.

Proof. Assume first that $0 \le v \le 1$ and that K is a given compact subset of U. Choose $v_j \in PSH \cap C^2$ near K, $v_j \searrow v_1$, $j \to +\infty$. We claim that

$$\lim_{\substack{j \to +\infty \\ 0 \le u \le 1}} \sup_{u \in PSH(U)} \int_K (v_j - v)(dd^c u)^n = 0.$$

Assuming this for a moment, choose k_j to be a fundamental sequence of compact subsets of U and let $\epsilon > 0$ be given.

Choose for every j, v_j such that

$$\sup_{\substack{u \in PSH(U) \\ 0 \leq u \leq 1}} \int_{k_j^0} (v_j - v)(dd'u)^n < \frac{\epsilon}{2^j}$$

and $v_{j+1} \leq v_j$ on k_j, $j \in \mathbb{N}$ and put $0_j = \{v_j - v > \frac{1}{j}\} \cap k_j^0$, $G_k = \bigcup_{j \geq k} 0_j$.

Then G_k is open, $d(G_k) \leq \epsilon \sum_{j=k}^{\infty} \frac{j}{2^j}$ and on $U - G_k$, $v_j \searrow v$ locally uniformly, $j \to +\infty$.

It remains to prove the claim. This is purely local so we can assume that $U = RB$ where $R > 1$, $K = \overline{B}$, $v = v_j$ outside B; Take $\tilde{u} \in PSH \cap C^2 x(RB)$, $\tilde{u} = $ const. near $\overline{B}$ and $\{u \neq \tilde{u}\} \subset L \subset\subset RB$. Then, for $k \geq j$, Stokes' formula gives

$$\int (v_j - v_k)(dd^c u)^n = -\int d(v_j - v_k) \wedge d^c u \wedge (dd^c u)^{n-1} =$$

$$= -\int d(v_j - v_k) \wedge d^c(u - \tilde{u}) \wedge (dd^c u)^{n-1}$$

since $\int d(v_j - v_k) \wedge d^c \tilde{u} \wedge (dd^c u)^{n-1} = 0$.

Since $(dd^c u)^{n-1}$ is a positive $(n-1, n-1)$-form we get

$$\left| \int d(v_j - v_k) \wedge d^c(u - \tilde{u}) \wedge (dd^c u)^{n-1} \right| \leq$$

$$\leq \left[\int d(v_j - v_k) \wedge d^c(v_j - v_k) \wedge (dd^c u)^{n-1} \right]^{\frac{1}{2}} \cdot$$

$$\cdot \left[\int d(u - \tilde{u}) \wedge d^c(u - \tilde{u}) \wedge (dd^c u)^{n-1} \right]^{\frac{1}{2}} =$$

$$= \left[\int (v_j - v_k) dd^c(v_j - v_K)(dd^c u)^{n-1} \right]^{\frac{1}{2}} \cdot$$

$$\cdot \int (u - \tilde{u}) dd^c(u - \tilde{u}) \wedge (dd^c u)^{n-1} \right)^{\frac{1}{2}} \leq$$

$$\leq \left(\int (v_j - v_k) dd^c(v_j + v_k) \wedge (dd^c u)^{n-1} \right)^{\frac{1}{2}} \cdot (2d(L))^{\frac{1}{2}}.$$

Continuing in this manner, we get an estimate of the form

$$\int (v_j - v_k)(dd^c u)^n \leq c[\int (v_j - v_k) dd^c (v_j + v_k)^n]^\alpha$$

where c is a constant and where $\alpha > 0$. Let $k \to +\infty$, then by Theorem V:2 $\int (v_j - v_k) dd^c (v_j + v_k)^n \to \int (v_j - v) dd^c (v_j + v)^n$. Let now $j \to +\infty$. Then again by Theorem V:2,

$$\int (v_j - v) dd^c (v_j + v)^n \to 0$$

which proves the claim.

c. <u>Dirichlet problem</u>

Theorem V:4. Let U be a strictly pseudoconvex and bounded domain in $\mathbb{C}^n$. If $h \in C(\partial U)$ and $f \in C(\overline{U})$, then there exists a unique plurisubharmonic function $u \in C(\overline{U})$ such that

$$\begin{cases} u = h \quad \text{on} \quad \partial U \\ (dd^c u)^n = f \quad \text{on} \quad U \end{cases}$$

d. <u>Continuity on increasing sequences</u>

Theorem V:5. Assume that $u_0, u_j : \in PSH \cap L^\infty(U)$, $j \in \mathbb{N}$, where $(u_j)_{j=1}^\infty$ is a non-decreasing sequence such that $\lim u_j = u_0$ (a.e.) when $j \to +\infty$.

Then $MA(\underbrace{u_j, \ldots, u_j}_{n \text{ times}}) \searrow MA(u_0, \ldots, u_0)$.

Proof. Let $\Theta \in C_0^\infty(U)$ be given. We have to prove that

$$\lim_{j \to +\infty} \int \Theta \, dd^c u_j \wedge \ldots \wedge dd^c u_j = \int \Theta \, dd^c u_0 \wedge \ldots \wedge dd^c u_0.$$

Take $\eta \in C_0^\infty(U)$ with $\eta = 1$ on supp Θ and $\Theta_1, \Theta_2 \in PSH \cap C^\infty$ with $\Theta = \Theta_1 - \Theta_2$. We then have

$$\int \Theta \, dd^c u_j \wedge \ldots \wedge dd^c u_j = \int u_j \, dd^c u_j \wedge \ldots \wedge dd^c u_j \wedge dd^c \Theta =$$

$$= \int \eta u_j \, dd^c u_j \wedge \ldots \wedge dd^c u_j \wedge (dd^c \Theta_1 - dd^c \Theta_2).$$

If we knew that

$$dd^c u_j \wedge \ldots \wedge dd^c u_j \wedge dd^c \Theta_1 \gtrless dd^c u_0 \wedge \ldots \wedge dd^c u_0 \wedge dd^c \Theta_1 \quad (*)$$

we would have for $k \leq j$,

$$\int \eta u_k \, dd^c u_j \wedge \ldots \wedge dd^c u_j \wedge dd^c \Theta_1 \leq \int \eta u_j \, dd^c u_j \wedge \ldots \wedge dd^c u_j \wedge dd^c \Theta_1 \leq$$

$$\leq \int \eta u_0 \, dd^c u_j \wedge \ldots \wedge dd^c u_j \wedge dd^c \Theta_1$$

by monotonicity, so since u_k are quasicontinuous:

$$\lim_{j \to +\infty} \int \eta u_k \, dd^c u_j \wedge \ldots \wedge dd^c u_j \wedge dd^c \Theta_1 =$$

$$= \int \eta u_k \, dd^c u_0 \wedge \ldots \wedge dd^c u_0 \wedge dd^c \Theta_1.$$

Thus $\int \eta u_k \, dd^c u_0 \wedge \ldots \wedge dd^c u_0 \wedge dd^c \Theta_1 \leq$

$$\varliminf_{j \to +\infty} \int \eta u_j \, dd^c u_j \wedge \ldots \wedge dd^c u_j \wedge dd^c \Theta_1 \leq \varlimsup_{j \to +\infty} \int \eta u_j \, dd^c u_j \wedge \ldots \wedge dd^c u_j \wedge dd^c \Theta_1 \leq$$

$$\varlimsup_{j \to +\infty} \int \eta u_0 \, dd^c u_j \wedge \ldots \wedge dd^c u_j \wedge dd^c \Theta_1 = \int \eta u_0 \, dd^c u_0 \wedge \ldots \wedge dd^c u_0 \wedge dd^c \Theta_1$$

where the last inequality follows from the fact that u_0 is quasicontinuous.

From Lemma V:1 below we get that

$$\lim_{k\to+\infty} \int \eta u_k \, dd^c u_0 \wedge \ldots \wedge dd^c u_0 \wedge dd^c \theta_1 = \int \eta u_0 \, dd^c u_0 \wedge \ldots \wedge dd^c u_0 \wedge dd^c \theta_1$$

and $\int u_j \, dd^c u_j \wedge \ldots \wedge dd^c u_j \wedge dd^c \theta_2$ can be handled in exactly the same way provided we know that

$$dd^c u_j \wedge \ldots \wedge dd^c u_j \wedge dd^c \Psi \rightharpoonup dd^c u_0 \wedge \ldots \wedge dd^c u_0 \wedge dd^c \Psi \quad \text{for every}$$
$\Psi \in C^\infty(U)$.

This we could do as above if we knew that

$$dd^c u_j \wedge \ldots \wedge dd^c u_j \wedge dd^c \Psi \wedge dd^c \gamma \rightharpoonup dd^c u_j \wedge \ldots \wedge dd^c u_j \wedge dd^c \Psi \wedge dd^c \gamma,$$
$\forall \Psi, \gamma \in C^\infty(U)$.

If we repeat this $n-1$ times we get a statement which is true. Thus (*) and Theorem V:5 is proved by the following Lemma.

Lemma V:1. If $|u_j| \leq$ const. $u_j \in PSH(U)$ $\forall j \in \mathbb{N}$, and if a.e. on U one has $u_j \nearrow u_0 \in PSH(U)$, then
$$u_j \, dd^c v^1 \wedge \ldots \wedge dd^c v^n \rightharpoonup u_0 \, dd^c v^1 \wedge \ldots \wedge dd^c v^n \quad \text{for all}$$
$v^i \in PSH(U) \cap L^\infty$, $1 \leq i \leq n$.

Proof. Fatous lemma gives that if μ is the weak limit of $u_j \, dd^c v^1 \wedge \ldots \wedge dd^c v^n$ then $\overline{\lim} \, u_j d^c v^1 \wedge \ldots \wedge dd^c v^n \geq \mu$. But $u_0 \geq \overline{\lim} \, u_i$ with equality almost everywhere with respect to the Lebesque measure so it is enough to prove that μ and $u_0 \, dd^c v^c \wedge \ldots \wedge dd^c v^n$ have (locally) the same mass. We can assume that $v^j = v^j_\varepsilon \in C^2$ outside some fix compact subset K of U where $v^j_\varepsilon \searrow v^j$, $\varepsilon \to 0$ on U.

If $\eta \in C_0^\infty(U)$, $\eta = 1$ near K then $\int \eta u_j dd^c v^1 \wedge \ldots \wedge dd^c(v^n - v_\varepsilon^n) =$

$\int u_j dd^c v^1 \wedge \ldots \wedge dd^c(v^n - v_\varepsilon^n) = \int (v^n - v_\varepsilon^n) dd^c v^1 \wedge \ldots \wedge dd^c v^{n-1} \wedge \ldots \wedge dd^c u_j \to 0$,

$\varepsilon \to 0$ uniformly in j since v^n is quasicontinuous. Thus

$\exists \varepsilon_1, \varepsilon_2, \ldots, \varepsilon_n$ such that $\int \eta u_j dd^c v^1 \wedge \ldots \wedge dd^c v_{\varepsilon_1}^n \geq$

$\geq -\varepsilon + \int \eta u_j dd^c v^1 \wedge \ldots \wedge dd^c v_\varepsilon^n \geq -n\varepsilon + \int \eta u_j dd^c v_{\varepsilon_n}^1 \wedge dd^c v_{\varepsilon_{n-1}}^2 \wedge \ldots \wedge dd^c v_{\varepsilon_1}^n$

where $0 < \varepsilon_n < \ldots < \varepsilon_1 < \varepsilon$ which gives $\lim_{j \to +\infty} \int \eta u_j dd^c v^1 \ldots \wedge dd^c v^n \geq$

$\int \eta u_0 dd^c v_{\varepsilon_n}^1 \wedge \ldots \wedge dd^c v_{\varepsilon_1}^n - \varepsilon n$. If we now let $\varepsilon \to 0$ and use the

quasicontinuity we get the desired conclusion.

e. <u>Comparison theorems</u>

Let U be and open and bounded subset of $\mathbb{C}^n$.

Lemma V:2. If $u, v \in PSH \cap L^\infty(U)$ and if $u = v$ near ∂U then

$$\int_U MA(u, \ldots, u) = \int_U MA(v, \ldots, v)$$

Proof. Given K, compact in U, choose $\chi \in C_0^\infty(U)$, $\chi \equiv 1$ near K, $u = v$ on $u \setminus K$. Then $\int_U \chi MA(u, \ldots, u) = \int \chi dd^c u \wedge \ldots \wedge dd^c u =$

$= \int u dd^c u \chi \wedge dd^c u \wedge \ldots \wedge dd^c u$ by Stokes formula. But $u = v$ on

supp $dd^c \chi$ so the right hand side equal $\int v dd^c \chi \wedge dd^c v \wedge \ldots dd^c v =$

$= \int \chi (dd^c v)^n = \int \chi MA(v, \ldots, v)$.

Lemma V:3. If $u, v \in PSH \cap L^\infty(U)$, $u \leq v$ on U and if $\lim_{\substack{z \to x \\ z \in U}} u(z) - v(z) = 0$

$\forall x \in \partial U$ then

$$\int_U MA(v, \ldots, v) \leq \int_U MA(u, \ldots, u).$$

Proof. Given $\varepsilon > 0$, put $v_\varepsilon = \sup(v-\varepsilon, u)$. Then $v_\varepsilon = u$ near ∂U so by Lemma V:2

$$\int_U MA(v_\varepsilon, \ldots, v_\varepsilon) = \int_U MA(u, \ldots, u).$$

Since $v_\varepsilon \nearrow v$ as $\varepsilon \searrow 0$, we have by Theorem V:5

$$MA(v_\varepsilon, \ldots, v_\varepsilon) \rightharpoonup MA(v, \ldots, v)$$

which proves the claim.

Theorem V:6. If $u, v \in PSH \cap L^\infty(U)$ and $\lim_{\substack{z \to \partial U \\ z \in U}} (u(z) - v(z)) \geq 0$ then

$$\int_{\{u < v\}} MA(v, \ldots, v) \leq \int_{\{u < v\}} MA(u, \ldots, u).$$

Furthermore, if $\lim_{\substack{z \to \partial U \\ z \in U}} u(z) - v(z) \geq \delta$ for some $\delta > 0$ then

$$\int_{\{u \leq v\}} MA(v, \ldots, v) \leq \int_{\{u \leq v\}} MA(u, \ldots, u).$$

Proof. We first note that the second statement follows from the first by considering u and $v + \varepsilon$ and letting ε decrease to zero. We also note that if u and v are also continuous, the first statement follows from Lemma V:3. To prove the theorem, it is no loss of generality to assume that $\overline{\{u \leq v\}} \subset U$ (consider otherwise $u + \varepsilon$ instead of u).

We first claim that if $(w_j)_{j=1}^\infty$ is a decreasing sequence of plurisubharmonic functions on U with $\lim_j w_j = w \in PSH \cap L^\infty(U)$

then

$$\int_{\{u<v\}} (dd^c w)^n \;\le\; \varliminf_{j\to+\infty} \int_{\{u<v\}} (dd^c w_j)^n$$

and

$$\int_{\{u\le v\}} (dd^c w)^n \;\ge\; \varlimsup_{j\to+\infty} \int_{\{u\le v\}} (dd^c w_j)^n.$$

For let $\delta>0$ be given. Since u and v are quasicontinuous there is an open set 0_δ with $\displaystyle\sup_j \int_{0_\delta}(dd^c w_j)^n<\delta$ and there are two continuous functions $\tilde{u}$ and $\tilde{v}$ such that $\{u\ne\tilde{u}\}\cup\{v\ne\tilde{v}\}\subset 0_\delta$. Therefore

$$\{u<v\}\subset\{\tilde{u}<\tilde{v}\}\cup\{u<v\}\cup 0_\delta \quad\text{and}$$

$$\{u\le v\} \;\le\; \{\tilde{u}\le\tilde{v}\}\cup 0_\delta \;\le\; \{u\le v\}\cup 0_\delta$$

$$\int_{\{u<v\}} (dd^c w)^n \;\le\; \varliminf \int_{\{\tilde{u}<\tilde{v}\}\cup 0_\delta} (dd^c w_j)^n \;\le\; \varliminf_{j\to+\infty} \int_{\{u<v\}} (dd^c w_j)^n + \delta$$

which gives the first part of the claim. Moreover

$$\varlimsup_{j\to+\infty} \int_{\{u\le v\}} (dd^c w_j)^n \;\le\; \varlimsup_{j\to+\infty} \int_{\{\tilde{u}\le\tilde{v}\}} (dd^c w_j)^n + \delta \;\le\; \int_{\{u\le v\}} (dd^c w)^n + 2\delta \quad\text{which}$$

gives the second part of the claim.

We can now finish the proof of the theorem. Choose u_δ and v_ε to the decreasing sequences of continuous plurisubharmonic functions defined in a neighborhood of $\overline{\{u<v_1\}}$ and such that $\overline{\{u<v_1\}}\subset U$.

Then $\displaystyle\int_{\{u<v\}} (dd^c v)^n \leq \binom{\text{the}}{\text{claim}} \leq \lim_{\varepsilon\to 0} \int_{\{u<v\}} (dd^c v_\varepsilon)^n \leq$

$\displaystyle\leq \lim_{\varepsilon\to 0}\lim_{\delta\to 0} \int_{\{u_\delta<u\}} (dd^c v_\varepsilon)^n \leq \lim_{\varepsilon\to 0}\lim_{\delta\to 0} \int_{\{u_\delta<v_\varepsilon\}} (dd^c v_\varepsilon)^n \leq \text{(Lemma V:3)} \leq$

$\displaystyle\leq \lim_{\varepsilon\to 0}\overline{\lim_{\delta\to 0}} \int_{\{u_\delta<v_\varepsilon\}} (dd^c v_\delta)^n \leq \overline{\lim_{\varepsilon\to 0}}\,\overline{\lim_{\delta\to 0}} \int_{\{u\leq v_\varepsilon\}} (dd^c u_\delta)^n \leq \text{(the claim)} \leq$

$\displaystyle\leq \overline{\lim_{\varepsilon\to 0}} \int_{\{u\leq v_\varepsilon\}} (dd^c u)^n = \int_{\{u\leq v\}} (dd^c u)^n.$

Hence $\displaystyle\int_{\{u+\eta<v\}} (dd^c v)^n \leq \int_{\{u+\eta\leq v\}} (dd^c u)^n$ for all small but positive η.

Let now η decrease to zero and we have the desired inequality.

Corollary V:1. If $u,v\in PSH\cap L^\infty(U)$ with $\displaystyle\lim_{\substack{z\to\partial U\\ z\in U}} u(z)-v(z)\geq 0$ and

if $MA(u,\ldots,u)\leq MA(v,\ldots,v)$ then $u\geq v$ on U.

Proof. Let $0\geq P\in PSH\cap L^\infty(U)$ such that $\displaystyle\int_V (dd^c P)^n>0$ for every

Borelset V in U with positive Lebesgue measure. If there is

a $z_0\in U$ with $u(z_0)<v(z_0)$ take $\eta>0$ so small that

$u(z_0)<v(z_0)+\eta P(z_0)$. Then the Lebesgue measure of

$T=\{z\in U;\ u<v+\eta P\}$ is strictly positive and so is $\displaystyle\int_T (dd^c P)^n$. By

Theorem V:6 we have that $\displaystyle\int_T (dd^c(v+\eta P))^n \leq \int_T (dd^c u)^n$ but the

right hand side is assumed to be smaller than $\displaystyle\int_T (dd^c v)^n$.

Hence $\displaystyle\int_T (dd^c v)^n+\eta^n\int_T (dd^c P)^n \leq \int_T (dd^c v)^n$ so $\displaystyle\int_T (dd^c P)^n = 0$

which is a contradiction.

Corollary V:2. If $u,v \in PSH \cap L^\infty(U)$, $\lim\limits_{\substack{z \to \partial U \\ z \in U}} u(z) - v(z) \geq 0$ and if

$$\int_{\{u<v\}} (dd^c u)^n = 0 \quad \text{then} \quad u \geq v \quad \text{on} \quad U.$$

Proof. If $\{u<v\} \neq \emptyset$ choose P as in the proof of Corollary V:1 and $\eta > 0$ so that

$$\{u < v + \eta P\} \neq \emptyset.$$

Then by Theorem V:6

$$\int_{\{u<v+\eta P\}} (dd^c v + \eta P)^n \leq \int_{\{u<v\}} (dd^c u)^n = 0.$$

Again $\displaystyle\int_{\{u<v+\eta P\}} (dd^c P)^n = 0$ which is a contradiction.

Theorem V:7. If K is compact in B and if

$$u_K = \sup\{\varphi \in PSH(RB);\ -1 \leq \varphi \leq 0;\ \varphi\big|_K = -1\} \quad \text{then} \quad \int_{CK} MA(u_K^*, \ldots, u_K^*) = 0.$$

Proof. If u_K is continuous, it follows from Theorem V:4 and Corollary V:1 that $\operatorname{supp} MA(u_K, \ldots, u_K) \subset K$.

If K is a given compact set, we can choose a decreasing sequence $\{K_s\}_{s=1}^\infty$ of compact sets decreasing to K where each u_{K_s} is continuous (cover K with finitely many balls). Then $\lim\limits_{s \to +\infty} u_{K_s} = u_K$ a.e. and an application of Theorem V:5 completes the proof.

Remark. If $K_j \searrow K$, $j \in \mathbb{N}$ is a decreasing sequence of compact subsets of B then $u^*_{K_j} \nearrow u^*_K$ so

$$d(K_j) = \int MA(u^*_{K_J}, \ldots, u^*_{K_j})$$

by Theorem V:6 and Theorem V:7.

Theorem V:5 now gives that

$$\lim_{j \to +\infty} d(K_j) = \int MA(u^*_{K_j}, \ldots, u^*_K) = d(K).$$

This together with Proposition V:1 proves that d is a capacity.

f. <u>Quasicontinuity in the system F,c</u>

Proposition V:2. If K is compact in B then

$$d(K) = \int_{RB} MA(u^*_K, \ldots, u^*_K) \quad \text{where} \quad u_K = \sup\{u \in PSH(RB); \; -1 \leq u \leq 0; \; u|_K = -1\}.$$

Proof. Given K compact in B we know from Theorem V:7 that

$$\int_K MA(u^*_K, \ldots, u^*_K) = \int_{RB} MA(u^*_K, \ldots, u^*_K).$$

Let $\varphi \in PSH(RB)$; $-1 < \varphi < 0$ be given. Since we want to estimate $\int_K MA(\varphi, \ldots, \varphi)$, it is no restriction to assume that
$$\lim_{z \to \gamma} \varphi(z) = 0, \quad \forall \gamma \in \partial RB.$$

Assume first that u_K is continuous. By Theorem V:6 we have

$$\int_K MA(\varphi,\ldots,\varphi) \leq \int_{\{u_K<\varphi\}} MA(u_K,\ldots,u_K) = \int_K MA(u_K,\ldots,u_K) =$$

$$= \int_B MA(u_K,\ldots,u_K).$$

If u_K is not continuous, choose $\{K_s\}_{s=1}^{\infty}$ to be a decreasing sequence of compact sets with intersection equal to K. Then $u_{K_s} \nearrow u_K$, $s \to +\infty$, a.e. so

$$\int_K MA(\varphi,\ldots,\varphi) = \lim_{s \to +\infty} \int_{K_s} MA(\varphi,\ldots,\varphi) \leq$$

$$\leq \lim_{s \to +\infty} \int_B MA(u_{K_s},\ldots,u_{K_s}) = \int_{RB} MA(u_K^*,\ldots,u_K^*)$$

where the last equality comes from Theorem V:5.

Corollary V:3. If 0 is an open subset of B then

$$d(O) = \int_{RB} MA(u_O,\ldots,u_O).$$

Proof. Choose an increasing sequence of compact sets $\{k_s\}_{s=1}^{\infty}$ with $\bigcup_{j=1}^{\infty} k_j = O$.

Then $d(k_s) = \int_{RB} MA(u_{k_s},\ldots,u_{k_s})$ by Proposition V:2. Thus

$$d(O) = \lim_{s \to +\infty} d(k_s) = \lim_{s \to +\infty} \int_{RB} MA(u_{k_s},\ldots,u_{k_s}) = \int_{RB} MA(u_O,\ldots,u_O)$$ by

Proposition V:1 and Theorem V:5.

We are now in position to prove the results stated in the introduction to this section. We first prove that 1) holds. Recall that F denotes the plurisuperharmonic functions on RB, $R>1$.

Given $u \in F$. By Theorem V:3 there is a decreasing sequence of open sets $\{O_s\}_{s=1}^{\infty}$ such that $\lim_{s \to +\infty} d(O_s)=0$ and such that $u|_{CO_s}$ is continuous. By Corollary V:3, $MA(u_{O_s},\ldots,u_{O_s}) \geq 0$, $s \to +\infty$.

It follows from Corollary V:1 that $\lim_{s \to +\infty} u_{O_s}(z)=0$ a.e. so, if necessary, by passing to a subsequence $\Psi = \sum_{s=1}^{\infty} u_{O_s} \in PSH(RB)$.

The restriction of u to $K_T = \{\Psi \geq -T\}$ is always continuous and $c(CK_T) \leq \frac{1}{T}\int_B \Psi d\delta \to 0$, $T \to +\infty$. This proves that every $u \in F$ is quasicontinuous with respect to c.

g.   <u>Condition 5) is fulfilled</u>

Let $0 \leq g$ be a bounded function. We wish to prove that $c(\{h_g > H_{h_g}\})=0$. Choose $h_i \in F: h_i \geq g$ so that $h_i \searrow H_{h_g}$ a.e. (δ), $i \to +\infty$ and put $A_{r_j} = \cap_i \{h_i > r_j > H_{h_g}\}$ where $\{r_j\}_{j=1}^{\infty}$ are the rational numbers in $(0,1)$. Then $\{h_g > H_{h_g}\} \subset \cup_{j=1}^{\infty} A_{r_j}$ so it is enough to prove that $c(A_{r_j})=0$, $\forall j \in \mathbb{N}$. Fix j and let k be a given compact subset of A_{r_j}; it is enough to prove that $c(k)=0$. Since $\frac{h_i}{r_j} \geq 1$ on k but $\frac{H_{h_g}}{r_j} < 1$ on A_{r_j} it is clear that $H_{h_g} < 1$ on k so $k \subset \cup_{s=1}^{\infty} \{H_{h_g} \leq 1-\frac{1}{s}\}$. Now $k_s = k \cap \{H_{h_k} \leq 1-\frac{1}{s}\}$ is compact and $H_{h_{k_s}} \leq 1-\frac{1}{s}$ on k_s. Assume that $\varphi \in F$; $\lim_{z \to \xi} \varphi(z)=0$,

$\forall \xi \in \partial RB$ and that $\varphi|_{k_s} \geq 1$. Then $H_{h_{k_s}} \leq (1-\frac{1}{s})\varphi$ on k_s. Furthermore, $MA(H_{h_{k_s}},\ldots,H_{h_{k_s}})=0$ on $RB \setminus k_s$ by Theorem V:7. Hence, by Corollary V:2 $H_{h_{k_s}} \leq (1-\frac{1}{s})\varphi$ on RB so we have that

$$H_{h_{k_s}} \leq (1-\frac{1}{s})h_{k_s} \quad \text{and so} \quad H_{h_{k_s}} \leq (1-\frac{1}{s})h_{k_s} \leq h_{k_s} = H_{h_{k_s}} \quad \text{a.e. } (d\delta).$$

It follows that $h_{k_s}=0$ a.e. $(d\delta)$ which proves the claim.

h.   Some consequences

We have now seen that in the plurisuperharmonic case, F and c satisfies conditions 1) and 5) respectively; here we list some consequences of this.

Theorem V:8. Let $\{u_j\}_{j \in I}$ be a family of plurisubharmonic functions, locally bounded above. Then there is a plurisubharmonic function Ψ such that

$$\{\sup_{j \in I} u_j < (\sup_{j \in I} u_j)^*\} \subset \{\Psi = -\infty\}.$$

Proof. Proposition III:3 gives 3) so the statement follows from Corollary III:1 and Lemma III:4.

Theorem V:9. The capacities c and d are outer regular and have the same zero sets. Furthermore

$$d(E) = \int (dd^c - h_{\chi_E})^n$$

for every Borel set E.

Proof. That c is outer regular follows from Proposition III:3 and Proposition III:1. By Lemma III:4 $c(E)=0 \Longleftrightarrow \exists u \in PSH$ such that $u=-\infty$ on E. Therefore $c(E)=0 \Rightarrow d(E)=0$. Assume for a moment that we have proved that

$$d(E) = \int (dd^c - h_{\chi_E})^n, \quad \text{for all Borel sets} \quad E.$$

If $d(E)=0$ then by Corollary V:I $-h_{\chi_E}=0$ a.e. so $c(E)=0$.

It remains to prove that $d(E)=\int (dd^c - h_{\chi_E})^n$ for all Borel sets E and we know this for E compact or open by Proposition V:2 and Corollary V:3. If E is a given Borelset then, since c is outer regular, we can find a decreasing sequence $(0_j)_{j=1}^{\infty}$ at open sets containing E such that $(-h_{\chi_E})^* = (\inf_j h_{\chi_{0_j}})^*$.

Hence $d(E) \leq \lim_j d(0_j) = \lim \int (dd^c - h_{\chi_{0_j}})^n = \int (dd^c - h_{\chi_E})^n$ by Theorem V:5. On the other hand, since

$$c(E) = \int h_{\chi_E} \, d\delta$$

is a capacity, we can find an increasing sequence $(K_j)_{j=1}^{\infty}$ of compact sets contained in E with $\lim_j c(K_j) = c(E)$ so

$$(-h_{\chi_{K_j}})^* \searrow (-h_{\chi_E})^*.$$

Then $d(E) \geq \lim_j d(K_j) = \lim \int (dd^c - h_{\chi_{K_j}}^*)^n = \int (dd^c (-h_{\chi_E})^*)^n$ by Theorem V:2.

Thus $d(E) = \int (dd^c(-h_{\chi_E})^*)^n$ which completes the proof of Theorem V:9.

Proposition V:3. The set function

$$G(E) = \inf_{\substack{-1<u<0 \\ u\in PSH(RB) \\ u=-1 \text{ on } E}} [\sup_{\substack{-1<v<0 \\ v\in PSH(RB)}} -\int u(dd^c v)^n]$$

is an outer regular capacity.

Proof. The results in this section show that Theorem III:3 applies.

Theorem V:10. Assume that $0 \geq u \in PSH \cap L^\infty(RB)$ when $R>1$. Then there are two sequences $\{u^p\}_{p=1}^\infty$ and $\{u_p\}_{p=1}^\infty$ of negative, continuous and plurisubharmonic functions on B such that

i) $u^p \searrow u$ on B

ii) $u^p + \sum_{t=p}^\infty u_t \nearrow u$ outside a set of c-capacity zero.

Proof. Theorem III:4.

In this section, we have seen the solution of the "coarse" problem (cf. Section III).

The pointwise ("fine") behaviour of $E \mapsto H_{\chi_E}(z)$, z fixed in B is not clear; we do not know if this is a capacity in Choquet's sense. The missing part is property ii): If $E_s \nearrow E$, $s \to +\infty$ we would like to know if $h_{\chi_{E_s}}(z) \nearrow h_{\chi_E}(z)$, $s \to +\infty$. What we have is the following theorem.

Theorem V:11. Assume that $K_s \nearrow K$, $s \to +\infty$ are compact sets of the unit ball. Then for every fixed $z \in B$, $\lim\limits_{s \to +\infty} h_{\chi_{K_s}}(z) = h_{\chi_K}(z)$.

Proof. Follows from Lemma III:1, since $E \mapsto c(\chi_E)$ is a capacity in Choquet´s sense.

Notes and references

General references:

Hörmander, L., An introduction to complex analysis in several variables. Van Nostrand, 1966.

Krantz, S.G., Function theory of several complex variables. Wiley-Interscience series, 1982.

Lelong, P., Plurisubharmonic functions and positive differential forms. Gordon and Breach, 1969.

Theorem V:8 as well as many other results in this section was first proved by **E. Bedford and B.A. Taylor** in: A new capacity for plurisubharmonic functions. Acta Math. Vol. 149 (1982).

Theorem V:4 is proved by the same authors in: The Dirichlet problem for the complex Monge-Ampère operator. Invent. Math. 37 (1976).

See also **L. Caffarelli, J.J. Kohn, L. Nirenberg and J. Spruck.**, The Dirichlet problem for non-linear second-order elliptic equations II. Complex Monge-Ampère, and uniformly elliptic equations. Communications on Pure and Applied Mathematics (1985).

S.-Y. Cheng and **S.T. Yau,** On the existence of a complete Kähler metric on non-compact complex manifolds and the regularity of Fefferman's equation. Communications on Pure and Applied Mathematics 33 (1980).

U. Cegrell, On the Dirichlet problem for the Monge-Ampère operator. Math. Z. 185 (1984).

The quasicontinuity has also been proved by **A. Sadullaev,** Rational approximation and pluripolar sets. Math. USSR Sbornik. Vol. 47 (1984). No. 1.

A stronger version of Theorem V:10 is proved in: **U. Cegrell,** Sums of continuous plurisubharmonic functions and the complex Monge-Ampère operator. Math. Z. 193 (1986).

In contradistinction to the subharmonic case, $E \mapsto h_{\chi_E}(x)$ is not strongly subadditive. This is shown by **Johan Thorbiörnson** in: A counterexample to the strong subadditivity of extremal plurisubharmonic functions. To appear in Mh. Math.

Theorem V:II was first proved in: **U. Cegrell,** Capacities and extremal plurisubharmonic functions on subsets of $\mathbb{C}^n$. Ark. Mat. 18 (1980).

A set E is called $\mathbb{C}^n$-polar if to every $z_0 \in E$ there is a ball $B(z_0,r)$ and $\varphi \in PSH(B(z_0,r))$ such that $\varphi \not\equiv -\infty$ and $E \cap B(z_0,r) \subset \{\varphi = -\infty\}$. It was shown by **B. Josefson**: On the equivalence between locally polar and globally polar sets for plurisubharmonic functions in $\mathbb{C}^n$. Ark. Mat. 16 (1978) that one always can take $\varphi \in PSH(\mathbb{C}^n)$.

For a stochastic point of view see **M. Fukushima** and **M. Okada**, On Dirichlet forms for plurisubharmonic functions. Acta Math. 159: 3-4, 1987.

VI Further Properties of the Monge-Ampère Operator

Let B be the unit ball in $\mathbb{C}^n$ and let P be the restriction to $\overline{B}$ of all negative plurisubharmonic functions on RB where $R>1$. We saw in Section V that $F=-P$ and δ, the Lebesgue measure on B, give rise to two natural capacities

$$d(E)=\sup\{\int_E (dd^c u)^n, -1<u<0, u\in P\} \quad \text{and} \quad c(E)=-\int u_E d\delta = \sup_{\mu\in M} \mu(E)$$

where $u_E(z)=\sup\{\varphi\in P;: \varphi\big|_E \leq -1\}(=-h_{\chi_E})$ and where M is defined to be the weak*-closed convex set of positive measures μ on B such that

$$\int \varphi d\delta \leq \int \varphi d\mu, \quad \forall \varphi\in P.$$

We can of course equally well consider the functionals

$$d(f)=\sup\{\int |f|(dd^c u)^n; \ u\in P: \ -1\leq u\leq 0\}$$

and

$$c(f)=\sup_{\mu\in M} \int |f| d\mu$$

where f is any bounded function on $\overline{B}$.

We do not know if $\{(dd^c u)^n; \ -1\leq u\leq 0, \ u\in P\}$ is compact or convex but if we define N to be the positive measures on $\overline{B}$ that are dominated by d (i.e. $0\leq \nu\in N \iff \nu(K)\leq d(\chi_K) \ \forall$ compacts K in $\overline{B}$) then obviously N is convex. But N is also compact f_0 if $\mu_j\in N$, $\mu_j \rightharpoonup \mu$ then, given $\varepsilon>0$ and K compact, choose an open subset A containing K such that $d(A)<d(K)+\varepsilon$. Then $\mu(K)\leq\mu(A)\leq \lim_{j\to+\infty}\mu_j(A)\leq d(A)\leq d(K)+\varepsilon$ by Lemma I:1, so $\mu\in N$.

Proposition VI:1. a) Every measure in M is absolutely continuous with respect to a measure in N.

b) There is a positive constant K such that $KN \subset M$.

Proof. a) Since N is weak*-compact and convex every Borel measure μ has a unique decomposition $\mu = \mu_1 + \mu_2$, where μ_1 is absolutely continuous with respect to a measure in N and μ_2 is carried by an F_σ-set E such that $\sup\{\nu(E), \nu \in N\} = 0$. If $\mu \in M$ it follows from Theorem V:9 that $\mu(E) = 0$ so $\mu_2 \equiv 0$ which proves the claim.

b) Follows from Proposition VI:2 below.

Proposition VI:2. There is a constant c such that

$$\int_{\overline{B}} - \varphi(dd^c u)^n \leq c \int_B - \varphi d\delta, \quad \forall \varphi \in P; \quad \forall u \in P, \quad -1 \leq u \leq 0.$$

Proof. Choose r, $1 < r < R$ and put $v_r(z) = \sup\left(\frac{\log|z|/r}{\log r}, -1\right)$. Then $v_n \geq u$ on $RB \setminus rB$ and $v_r \leq u$ on B. If we put $u_1 = \sup(u, v_r)$ we get

$$\int_{\overline{B}} - \varphi(dd^c u)^n \leq \int_{rB} - \varphi(dd^c u_1)^n =$$

$$= \int_{rB} - \varphi dd^c v_r \wedge (dd^c u_1)^{n-1} + \int - \varphi dd^c(u_1 - v_r) \wedge (dd^c u_1)^{n-1} =$$

$$= \int_{rB} - \varphi dd^c v_r \wedge (dd^c u_1)^{n-1} - \int (u_1 - v_r) dd^c \varphi \wedge (dd^c u_1)^{n-1} \leq$$

$$\leq \int_{rB} - \varphi dd^c v_r \wedge (dd^c u_1)^{n-1} \leq \text{ (repeat n-1 times) } \leq$$

$$\leq \int_{rB} - \varphi(dd^c v_r)^n = c_r \int_{|w|=r} - \varphi(w) d\sigma(w)$$

where σ is the surface measure on the boundary of rB in $\mathbb{C}^n$ and where c_r is a constant, only depending on r and n. Now, since φ is subharmonic we know

$$\int_B \varphi \, d\delta \leq \frac{1}{r^{2n}} \int_{rB} \varphi \, d\delta \leq \frac{1}{r^{2n}} \int \varphi \, d\sigma .$$

Therefore

$$c_r \int - \varphi \, d\delta \leq r_r r^{2n} \int_B - \varphi \, d\delta$$

which proves Proposition VI:2.

Note that if K is a compact subset of B then of course $d(\chi_K) \leq d(u_K)$ and we can have strict inequality.

For let

$$K = \{ z \in \mathbb{C}^2 ; \ |z| \leq \tfrac{1}{4} \}$$

and

$$L = \{ z \in \mathbb{C}^2 ; \ |z| \leq \tfrac{3}{4} \} .$$

Then

$$u_K = \max \{ \frac{\log |z|}{\log 4} , \ -1 \} ,$$

$$u_L = \max \{ \frac{\log |z|}{\log 4/3} , \ -1 \}$$

and

$$d(\chi_K) = \int (dd^c h_K)^2 .$$

Now $\quad d(u_K) \geq \int u_K (dd^c h_L)^2 = \int u_L dd^c u_L \quad dd^c u_K =$

$$= \frac{1}{\log 4 \ \log 4/3} \int u_L dd^c [\max(\log|z|,-\log 4)] \ dd^c [\max(\log|z|,-\log 4/3)] =$$

$$= \frac{1}{\log 4 \ \log 4/3} \int u_L dd^c [\max(\log|z|,-\log 4/3)]^2 =$$

$$= \frac{1}{\log 4 \ \log 4/3} \int dd^c [\max(\log|z|,-\log 4)]^2 >$$

$$> \frac{1}{(\log 4)^2} \int dd^c [\max(\log|z|,-\log 4)]^2 = \int (dd^c u_K)^2 = d(u_K).$$

Note that in view of the results in Section V,

$$E \rightarrow d(\chi_E)$$

and

$$E \rightarrow d(u_E)$$

are outer regular capacities.

We now turn to the problem of estimating the Monge-Ampère mass of plurisubharmonic functions defined outside a compact set.

Let Ω be an open and bounded pseudoconvex set in $\mathbb{C}^n$, $n \geq 2$ and K a compact subset of Ω so that $\Omega \backslash K$ is connected. By Hartogs extension theorem, every analytic function on $\Omega \backslash K$ extends to Ω. When it comes to plurisubharmonic functions the situation is different.

Proposition VI:3. Assume that $\varphi \in PSH \cap L^\infty_{loc}(\Omega \backslash K)$ where K is a removable singularity set for the plurisubharmonic functions. Then $\int_{\Omega' \backslash K} (dd^c \varphi)^n < +\infty$ when $K \subset\subset \Omega' \subset\subset \Omega$.

Proof. From Section V we know that $(dd^c\varphi)^n$ is a well-defined positive measure on $\Omega\setminus K$. Since φ extends to a plurisubharmonic function on Ω, we can choose a sequence $\varphi_j\in PSH\cap C^\infty$ near $\overline{\Omega'}$ such that $\varphi_j\searrow\varphi$, $j\to+\infty$.

Let $W=\{z\in\Omega'; d(z,K)>\frac{1}{2}d(\partial\Omega',K)\}$. Then $\delta=\inf\limits_{z\in W}\varphi(z)>-\infty$ and if we put $\psi_j=\sup(\varphi_j,\delta)$ then $\psi_j=\varphi_j$ on W.

So $\int\limits_{\Omega'}(dd^c\psi_j)^n=\int\limits_{\Omega'}(dd^c\varphi_j)^n$ by Stokes' theorem.

Therefore $\int\limits_{\Omega'\setminus K}(dd^c\varphi)^n\leq\lim\limits_{j\to+\infty}\int\limits_{\Omega'}(dd^c\varphi_j)=\lim\limits_{j\to+\infty}\int\limits_{\Omega'}(dd^c\psi_j)=$

$=\int\limits_{\Omega'}(dd^c\sup(\varphi,\delta))^n<+\infty$ since $\sup(\varphi,\delta)\in PSH\cap L^\infty_{loc}(\Omega)$.

Theorem VI:1. Let $\psi\in PSH\cap C^2(\Omega)$ and assume that $\Omega'=\{\psi<1\}$ and that $K=\{\psi\leq s\}$ for an s, $0<s<1$. If $\varphi\in PSH\cap L^\infty_{loc}(\Omega\setminus K)$ then

$$\int\limits_{\{s<\psi<1\}}|\varphi|(dd^c\varphi)^{n-1}\wedge dd^c\psi<+\infty$$

and

$$\int\limits_{\{s<\psi<1\}}(\psi-s)(dd^c\varphi)^n<+\infty.$$

Proof. Since Ω is pseudoconvex and ψ plurisubharmonic, there is to every $z_0\in\{s<\psi<1\}$ an analytic functions such that $f(z)\neq f(z_0)$, $\forall z\in K$. If we restrict φ to $\{f(z)=f(z_0)\}$ and apply the maximum principle, we conclude that φ is uniformly bounded above on $\overline{\Omega'}\setminus K$. It is therefore no restriction to assume that $\varphi \leq 0$.

Let $s<r<1$. Then we have by Stokes' theorem

$$\int_{\{r<\psi<1\}} -\varphi(dd^c\psi)\wedge(dd^c\varphi)^{n-1} + \int_{\{r<\psi<1\}} (\psi-r)(dd^c\varphi)^n \leq$$

$$\leq \lim_{t \to r} \int_{\{r<\psi<1\}} -\varphi dd^c\max(\psi,t)\wedge(dd^c\varphi)^{n-1} + \int_{\{r<\psi<1\}} (\psi-r)(dd^c\varphi)^n =$$

$$= \lim_{t \to r}\left[\int_{\{\psi=1\}} -\varphi d^c\max(\psi,t)\wedge(dd^c\varphi)^{n-1} + \right.$$

$$+ \int_{\{r<\psi<1\}} d\max(\psi,t)\wedge d^c\varphi\wedge(dd^c\varphi)^{n-1} + \left. \int_{\{r<\psi<1\}} (\psi-r)(dd^c\varphi)^n \right] =$$

$$= \lim_{t \to r}\left[\int_{\{\psi=1\}} -\varphi d^c\max(\psi,t)\wedge(dd^c\varphi)^{n-1} + \right.$$

$$+ \int_{\{\psi=1\}} (\max(\psi,t)-r)d^c\varphi\wedge(dd^c\varphi)^{n-1} -$$

$$- \int_{\{\psi=r\}} (\max(\psi,t)-r)d^c\varphi\wedge(dd^c\varphi)^{n-1} +$$

$$+ \left. \int_{\{r<\psi<1\}} (r-\max(\psi,t))(dd^c\varphi)^n \right] \leq \int_{\{\psi=1\}} -\varphi d^c\psi\wedge(dd^c\varphi)^{n-1} +$$

$$+ \int_{\{\psi=1\}} (\psi-r)d^c\varphi\wedge(dd^c\varphi)^{n-1}.$$

Since the right hand side is uniformly bounded in $r,s<r<1$ and since $\varphi \leq 0$ we get

$$\int_{\{s<\psi<1\}} |\varphi| \, dd^c\psi \wedge (dd^c\varphi)^{n-1} + \int_{\{s<\psi<1\}} (\psi-s)(dd^c\varphi)^n < +\infty$$

where each number is non-negative.

An example

The following example shows that the convergence factor $\psi-s$ really is needed in the theorem. There are functions φ and ψ such that

$$\int_{\{s<\psi<1\}} (dd^c\varphi)^n = +\infty$$

where φ can be taken to be plurisubharmonic and bounded.

Define $u(z) = -\sqrt{|z|^2 - \frac{1}{2}}$; then u is subharmonic on $\{z\in\mathbb{C}; \frac{1}{2} < |z|^2 < 1\}$, $-\frac{1}{\sqrt{2}} < u < 0$ and $\Delta u = \dfrac{1 - |z|^2}{(|z|^2 - \frac{1}{2})^{3/2}}$.

Define $\psi(z,w) = |z|^2 + \frac{2}{3}|w|^8$ and let $0 < \eta < \frac{1}{20}$ and put

$$\varphi_\eta(z,w) = \max\left[-\frac{1}{2}\sqrt{(1+\eta)^2|z|^2 - \frac{1}{2}} + |w|^2, \ \ |w|^2 + (1+\eta)|w|^4 - \sqrt{\frac{3}{2}(\eta + \frac{\eta^2}{2})} \right].$$

Then φ_η is plurisubharmonic on $\{\frac{1}{2} < \psi < (\frac{20}{21})^2\}$ for if $(1+\eta)^2|z|^2 \le \frac{1}{2}$ we have, when $\psi(z,w) > \frac{1}{2}$

$$\frac{2}{3}|w|^8 > \frac{1}{2} - |z|^2 > \frac{1}{2}(1-(1+\eta)^{-2}) = \frac{1}{2}\frac{2\eta+\eta^2}{(1+\eta)^2} = \frac{\eta + \frac{\eta^2}{2}}{(1+\eta)^2} \quad \text{so}$$

$$(1+\eta)|w|^4 > \sqrt{\frac{3}{2}(\eta + \frac{\eta^2}{2})} \ .$$

$$- 63 -$$

Assume now that $\frac{1}{2} < |z|^2 + \frac{2}{3}|w|^8 < (\frac{20}{21})^2$ and that

$$\sqrt{\frac{3}{5}} \ \frac{\sqrt{\eta + \frac{\eta^2}{2}}}{1 + \eta} < |w|^4 < \sqrt{\frac{2}{3}} \ \frac{39}{40} \ \frac{\sqrt{\eta + \frac{\eta^2}{2}}}{1 + \eta} \ .$$

Then $\varphi_\eta = -\frac{1}{2} \sqrt{(1+\eta)^2 |z|^2 - \frac{1}{2}} + |w|^2$ if

$$- \frac{1}{2} \sqrt{(1+\eta)^2 |z|^2 - \frac{1}{2}} > (1+\eta)|w|^4 - \sqrt{\frac{3}{2}(\eta + \frac{\eta^2}{2})}$$

$$\Leftrightarrow (1+\eta)^2 |z|^2 < \frac{1}{2} + 4 \left(\sqrt{\frac{3}{2}(\eta + \frac{\eta^2}{2})} - (1+\eta)|w|^4 \right)^2 \quad \text{which holds if}$$

$$(\frac{1}{2} - \frac{2}{3}|w|^8)(1+\eta)^2 < (1+\eta)^2 |z|^2 < \frac{1}{2} + 4 \left(\sqrt{\frac{3}{2}(\eta + \frac{\eta^2}{2})} - (1+\eta)|w|^4 \right)^2$$

which is a well defined domain if

$$\eta + \frac{\eta^2}{2} - \frac{2}{3}|w|^8(1+\eta)^2 < 6(\eta + \frac{\eta^2}{2}) - 8\sqrt{\frac{3}{2}(\eta + \frac{\eta^2}{2})}(1+\eta)|w|^4 + 4(1+\eta)^2|w|^8$$

$$\Leftrightarrow$$

$$0 < 5(\eta + \frac{\eta^2}{2}) + \frac{14}{3}(1+\eta)^2 |w|^8 - 8\sqrt{\frac{3}{2}} \sqrt{\eta + \frac{\eta^2}{2}}(1+\eta)|w|^4$$

which holds with $|w|^4$ in the interal above - the right hand
side is then strictly larger than

$$5(\eta + \frac{\eta^2}{2}) + \frac{14}{3}(1+\eta)^2 \ \frac{3}{5} \ \frac{\eta + \frac{\eta^2}{2}}{(1+\eta)^2} - 8 \cdot \frac{39}{40}(\eta + \frac{\eta^2}{2}) = 0 \ .$$

Therefore
$$\int\limits_{\frac{1}{2} < |z|^2 + \frac{2}{3}|w|^8 < (\frac{20}{21})^2} (dd^c \varphi_\eta)^2 \geq c \int\limits_{A_\eta} \frac{dz \ dw}{((1+\eta)^2 |z|^2 - \frac{1}{2})^{3/2}}$$

where

$$A_\eta = \begin{cases} (\frac{1}{2} - \frac{2}{3}|w|^8)(1+\eta)^2 < (1+\eta)^2|z|^2 < \frac{1}{2}+4\left(\sqrt{\frac{3}{2}(\eta+\frac{\eta^2}{2})} - (1+\eta)|w|^4\right)^2 \\[2ex] \dfrac{\sqrt{\eta + \frac{\eta^2}{2}}}{1+\eta}\,\alpha < |w|^4 < \beta\,\dfrac{\sqrt{\eta + \frac{\eta^2}{2}}}{1+\eta} \end{cases}$$

and where we have chosen $\sqrt{\frac{3}{5}} < \alpha < \beta < \frac{39}{40}\sqrt{\frac{2}{3}}$ and where c is a strictly positive constant. The right hand side is not smaller than

$$c\{(\alpha^2-\frac{3}{5}) + \frac{39}{40}\sqrt{\frac{2}{3}} - \beta\}\int_\Omega \frac{(\eta + \frac{\eta^2}{2})dw}{2^3(\sqrt{\frac{3}{2}(\eta + \frac{\eta^2}{2})} - (1+\eta)|w|^4)^3} \geq$$

$$\geq d(\eta + \frac{\eta^2}{2})^{1-\frac{3}{2}-\frac{1}{4}} = \frac{d}{(\eta + \frac{\eta^2}{2})^{1/4}} \to +\infty, \quad \eta\to 0 \quad (d \text{ positive constant})$$

where $\quad \Omega = \left\{ \dfrac{\sqrt{\frac{3}{5}(\eta + \frac{\eta^2}{2})}}{1+\eta} < |w|^4 < \sqrt{\frac{2}{3}}\,\frac{39}{40}\,\dfrac{\sqrt{\eta + \frac{\eta^2}{2}}}{1+\eta} \right\}.$

Furthermore, $-\frac{1}{2} < \varphi_\eta \leq 2$ so if we put

$$\varphi = \sum_{j=20}^\infty \frac{1}{j^2}\,\varphi_{1/j^4}$$

then $\varphi \in PSH \cap L^\infty\left(\frac{1}{2} < |z|^2 + \frac{2}{3}|w|^8 < \left(\frac{20^4}{20^4+1}\right)^2\right)$

and $\displaystyle\int_{\frac{1}{2}<|z|^2+\frac{2}{3}|w|^8<(\frac{20}{21})^2} (dd^c\varphi)^2 = +\infty.$

Remark. It is possible to modify the φ_η:s to get the function φ of class C^∞ on $\Omega\backslash\{w=0\}$. We do not know if one can modify φ to be C^∞ on Ω but still have $\int\limits_{\Omega\backslash K}(dd^c\varphi)^2=+\infty$.

Notes and references

For the decomposition of measures used in the proof of Proposition VI:1 a) see Section XI.

Proposition V:2 was first proved by **Demailly, J.-P.**, Mesures de Monge-Ampère et caracterization geometrique des varietes algebriques affines. Bulletin de la Société Mathématiques de France, Memoire 19, 113 (1985), 1-125.

Proposition VI:3 in the case K equal to one point was proved by **Griffith, P.**, Two theorems on extension of holomorphic mappings. Inv. Math. 14 (1971), 27-62.

Theorem VI:1 in the plurisubharmonic and C^2 case was proved by **Fornaess, J.E.** and **Sibony, N.**, Plurisubharmonic functions on ring domains. Krantz, S.G., Ed., Complex analysis. Seminar University Park PA, 1986. Springer Lecture Notes in Mathematics. Vol. 1268.

Proposition VI:1 and Theorem VI:1 will appear in **Cegrell, U.**, Plurisubharmonic functions outside compact sets. Proc. AMS.

The above example shows that there are functions that can be subextended and still have unbounded Monge-Ampère mass near a compact set.

In smooth domains, there are plurisubharmonic functions that cannot be subextended: **Bedford, E.** and **Taylor, B.A.**, Smooth plurisubharmonic functions with no subextensions. Manuscript 1986.

VII Green's Function

An open subset D of $\mathbb{C}^n$ is called pseudoconvex if $-\log d(z,CD)$ is plurisubharmonic. $(d(z,CD)=\inf_{w\in CD}|z-w|)$.

If there is a continuous and plurisubharmonic function $\varphi<0$ on D so that

$$\{z\in D;\ \varphi(z)<\alpha\}\subset\subset D,\ \forall\alpha<0$$

then D is said to be hyperconvex.

If there is a C^2-function $\rho:\mathbb{C}^n\to\mathbb{R}$, grad $\rho\neq 0$ on D, $D=\{\rho<0\}$

$$\sum_{i,j=1}^{n}\frac{\partial^2\rho}{\partial z_i\partial\overline{z}_j}(P)w_i\overline{w}_j\geq c|w|^2,\ \forall P\in\partial D,\ \forall w\in\mathbb{C}^n.$$

for some $c>0$ then D is said to be strictly pseudoconvex.

Definition. Let D be an open subset of $\mathbb{C}^n$, $n\geq 1$ and assume that $\psi\in SH(D\times D)$. Then ψ is called 2-plurisubharmonic (2-PSH) on $D\times D$ if

$$\Omega\ni z\to\psi(z,w)\in PSH(D),\ \forall w\in D$$

$$\Omega\ni w\to\psi(z,w)\in PSH(D),\ \forall z\in D.$$

Definition. Green's function relatively to D is

$$W(z,w)=\sup\{u(z,w)\in 2PSH(D\times D);\ u\leq 0$$

$$u(z,w)\leq\log|z-w|-\log\max[d(z,CD),\ d(w,CD)]\}.$$

Lemma VII:11. If $u \in PSH(D)$, $u \leq 0$, $w_0 \in D$ and if $u(z) - \log|z-w_0|$ is bounded above near w_0 then

$$u(z) \leq \log|z-w_0| - \log\, d(z,CD), \quad \forall z \in D.$$

Proof. Consider for $\tau \in \mathbb{C}$ and $z \in D$

$$V_z = \{\tau;\ \tau(z-w_0)+w_0 \in D\}.$$

Then $1 \in V_z$ and since

$$g(\tau) = u(\tau(z-w_0)+w_0) - \log|\tau|\ |z-w_0|$$

is subharmonic on V_z we get

$$g(1) \leq \sup_{\tau \in V_z} g(\tau) \leq \sup_{\tau \in \partial V_z}\ -\log|\tau(z-w_0)| =$$

$$= -\log \inf_{\tau \in \partial V_z} |\tau(z-w_0)| \leq -\log\, d(w_0,CD).$$

Therefore $u(z) \leq \log|z-w_0| - \log\, d(w_0,CD)$, $z \in D$.

Corollary VII:1. If $u \leq 0$, $u(z,w) - \log|z-w|$ locally bounded near the diagonal $\Delta \subset D \times D$ and if $u \in 2\text{-}PSH(D \times D)$ then

$$u(z,w) \leq \log|z-w| - \log\, \max[d(z,CD),\ d(w,CD)].$$

Corollary VII:2. If $f \in H^\infty(D)$, $|f| \leq 1$, $f \not\equiv const.$ then

$$v_f(z,w) = \log\left|\frac{f(z)-f(w)}{1-f(z)\overline{f}(w)}\right| \leq W(z,w).$$

Proof. It is clear that $v_f \in 2\text{-PSH}(D \times D)$ (but v_f is not $\text{PSH}(D \times D)$!), $v_f \leq 0$. Furthermore, $v_f(z,w) - \log|z-w|$ is locally bounded above on Δ so by Corollary VII:1, $v_f \leq W$.

Proposition VII:1. 1) $W(z,w) = W(w,z)$

2) $W \in 2\text{-PSH}$

3) If furthermore D is strictly pseudoconvex then

$$\lim_{z \to \partial D} W(z,w) = 0, \quad \forall w \in D.$$

Proof. 1) Since $\log|z-w| - \log \max[d(z,CD), d(w,CD)]$ is symmetric so is W.

2) Denote by W^* the smallest upper semicontinuous majorant of W. Then W is subharmonic on $D \times D$ and we claim that W^* is $2\text{-PSH}(D \times D)$. If we take φ to be a function only depending on $|z|$ and consider $W_{\varepsilon,\delta}(z,w) = \int W(\xi,n)\varphi(\frac{z-\xi}{\varepsilon})\varphi(\frac{w-n}{\delta})$ then $W_{\varepsilon,\delta} \searrow W^*$ when $\varepsilon,\delta \searrow 0$ and $W_{\varepsilon,\delta}$ is 2-PSH. Furthermore $W^*(z,w) \leq \log|z-w| - \log \max[d(z,CD), d(w,CD)]$ since the right hand side is upper semicontinuous. Therefore $W = W^*$ which proves 2).

3) If D is strictly pseudoconvex it is well known that to every $P \in \partial D$ there is a function f_p, analytic on D and continuous on $\overline{D}$ such that $f(p)=1$ and $|f|<1$ on D.

By Corollary VII:2
$$v_{f_p}(z,w) = \log\left|\frac{f(z)-f(w)}{1-f(z)\overline{f}(w)}\right| \leq W(z,w) \quad \text{so therefore}$$

$$0 \geq \varlimsup_{z \to p} W(z,w) \geq \varliminf_{z \to p} W(z,p) \geq \lim_{z \to p} v_{f_p}(z,w) = 0 \quad \text{which completes}$$

the proof of Proposition VII:1.

Theorem VII:1. Let $f: D \to D'$ be a holomorphic map between two open sets $D \subset \mathbb{C}^n$, $D' \subset \mathbb{C}^n$. Then

$$W_{D'}(f(z),f(w)) \leq W_D(z,w).$$

Proof. It is clear that $W_{D'}(f(z),f(w))$ is negative and 2-PSH(DxD).

Furthermore $W_{D'}(f(z),f(w)) \leq \log|f(z)-f(w)| -$
$-\log \max[d(f(z),CD'),d(f(w),CD')] = \log|z-w| + \log \dfrac{|f(z)-f(w)|}{|z-w|} -$
$-\log \max[d(f(z),CD'), d(f(w),CD')]$ so $W_{D'}(f(z),f(w)) - \log(z-w)$
is locally bounded above near Δ. Therefore $W_{D'}(f(z),f(w)) \leq$
$\leq W_D(z,w)$.

Klimek and Demailly have studied the following function.
$u(z,w) = u_D(z,w) = \sup\{u(z) \in PSH(D); u \leq 0, u(z)-\log|z-w| \text{ bounded}$
above near $w\}$.

It is clear that $W \leq u$.

Proposition VII:2. Assume that D is strictly pseudoconvex. The following conditions are equivalent.

i) $\displaystyle \lim_{t \to -\infty} \int_D (dd^c_\xi \max(W(z,\xi),t))^n = (2\pi)^n, \quad \forall z \in D$

ii) $W = u$

iii) $u(z,w) = u(w,z), \quad \forall (z,w) \in DxD$

iv) $D \ni w \mapsto u(z,w) \in PSH(D), \quad \forall z \in D.$

Proof. ii) $\Rightarrow$ iii) follows from Proposition VII:1.

iii) $\Rightarrow$ iv) since $z \to u(z,w) \in PSH(D)$, $\forall w \in D$.

iv) $\Rightarrow$ ii). It follows from Corollary VII:1 that
$u(z,w) \leq \log|z-w| - \log \max[d(z,CD), d(CD)]$ so $u \leq W$. On the other
hand, it is always true that $u \geq W$.

ii) $\Rightarrow$ i) $(dd^c u(z,w))^n = 0$ outside w. Therefore

$$\lim_{t \to -\infty} \int_D (dd^c(\max u(z,w),t))^n = \lim_{t \to -\infty} \int_D (dd^c \max(\log|z-w|,t))^n = (2\pi)^n.$$

i) $\Rightarrow$ ii). Fix $w \in D$. Since $\max(W(z,w),t) \leq \max(u(z,w),t)$ and
since both functions have boundary values zero, it follows from
Lemma V:3 that

$$\int_D (dd^c \max(W(z,w),t))^n \geq \int_D (dd^c \max(u(z,w),t))^n = (2\pi)^n.$$

Therefore i) means that $(dd^c W(z,w))^n = 0$ for $z \neq w$ because,
since $W(z,w) - \log|z-w|$ is bounded near $z=w$,
$\lim_{t \to -\infty} \int_E (dd^c \max(W(z,w),t))^n = (2\pi)^n$ for every neighborhood E of
w. Then $(1-\varepsilon)W(z,w)$ is equal to zero at ∂D, larger than
$u(z,w)$ near w. Therefore, Corollary V:1 gives that
$(1-\varepsilon)W(z,w) \geq u(z,w)$ for every $\varepsilon > 0$ which completes the proof
of Proposition VII:2.

Definition. Let D be a domain in $\mathbb{C}^n$. If $(z,w) \in D \times D$ we
define the **Caratheodory distance**

$$C_D(z,w) = \sup_F \{\rho(F(z),F(w)); \ F: D \to U, \ F \text{ holomorphic}\}$$

and the **Kobayashi distance**

$$K_D(z,w)=\inf\{\sum_{j=1}^{m}\delta_D(z_j,z_{j-1});\ z_0=z,\ z_m=w\}$$

where $\delta_D(z,w)=\inf\{\rho(\xi,\eta);\ \exists f:U\to D$ with $f(\xi)=z,\ f(\eta)=w,$ f holomorphic$\}$ where U is the unit disc in $\mathbb{C}$ and where ρ is the hyperbolic distance on U.

Proposition VII:3. $\log\tanh C(z,w)\leq W(z,w)\leq u(z,w)\leq\log\tanh\delta(z,w).$ If D is convex, then equality holds.

Proof. We have that

$$\log\tanh C(z,w)=\sup\{\log\left|\frac{f(z)-f(w)}{1-f(z)\overline{f}(w)}\right|;\ f:D\to U,\ f\ \text{holomorphic}\}$$

so by Corollary VII:2 the first inequality follows. The second is clear and the third follows from Theorem VII:1 since $u_U=\log\tanh(\rho)$. If D is convex, it is a result of Lempert that $C=\delta$.

When $n=1$, we have for every compact set $K\subset U$, the unit disc

$$u_K^*(z) = \int W(z,\xi)d\mu(\xi),\ z\in U$$

where μ is a uniquely determined positive measure on K.

When $n>1$, this is no longer true since there are compact pluripolar sets supporting positive measures such that $\int W(z,\xi)d\mu(\xi)$ is bounded on D.

Notes and references

The function u is introduced by **M. Klimek** in the article
Extremal plurisubharmonic functions and pseudodistances. Bull.
Soc. Math. France 113 (1985).

In Mesures de Monge-Ampère et mesures pluriharmonique by
J.-P. Demailly, Math. Z. 194 (1987), 519-564, this study is
continued, including a representation formula and explicit
examples.

That $C_D = K_D$ for convex sets D was proved by **L. Lempert**
in Analysis Mathematica 8 (1982).

VIII The Global Extremal Function

Denote by L the class of plurisubharmonic functions f on $\mathbb{C}^n$ such that

$$f(z) \leq a_f + \log^+|z|, \quad z \in \mathbb{C}^n$$

where a_f is a constant (depending on f). If $E \subset \mathbb{C}^n$ we define

$$V_E(z) = \sup\{f(z); \; f \in L; \; f \leq 0 \quad \text{on} \quad E\}$$

and V_E^* to be the smallest upper semicontinuous majorant of V_E.

Proposition VIII:1. $V_E^* \in L \iff E$ is not pluripolar.

If $V_E^* \in L$ then the function is called the global extremal plurisubharmonic function relatively to E.

Theorem VIII:1. Assume that E is a relatively compact subset of the unit ball. Then $(dd^c V_E^*)^n$ and $(dd^c h_E^*)^n$ are mutually absolutely continuous. (See VI for the definition of h_E.)

Let $u \in L$ and define

$$\gamma_r(u) = \int\limits_{|w|=1} u(rw)\,d\sigma(w) - \log r$$

where σ is the normalized Lebesgue measure on the unit sphere. Then $\gamma_{e^r}(u)$ is convex and bounded at $+\infty$ so $\gamma_r(u)$ is decreasing and we denote by $\gamma(u)$ the limit $\lim\limits_{r \to +\infty} \gamma_r(u)$. If furthermore $u \geq \log^+|z|$ then $\gamma(u) \geq 0$.

Lemma VIII:1. There is a constant $C_n = (2\pi)^n$ so that

$$C_n = \int (dd^c u)^{n-p} \wedge (dd^c v)^p, \quad \forall \ 0 \leq p \leq n, \ \forall \ \log^+|z| \leq u, v \in L.$$

Proposition VIII:2. If Θ is a d-closed $(n-1, n-1)$-form then

$$\int_{|z|<R} u dd^c \log^+|z| \wedge \Theta = \int_{|z|<R} \log^+|z| dd^c u \wedge \Theta +$$

$$+ \int_{|z|=R} u d^c \log^+|z| \wedge \Theta - \log R \int_{|z|=R} d^c u \wedge \Theta.$$

Proof. Stokes' formula.

Theorem VIII:2. If $\log^+|z| \leq u \in L$ then

$$\gamma(u) = \int_{|u|=1} u(w) d\sigma(w) - \frac{1}{C^n} \int_{\mathbb{C}^n} \log^+|z| dd^c u \wedge (dd^c \log^+|z|)^{n-1}.$$

Proof. Take Θ to be $(dd^c \log^+|z|)^{n-1}$ in Proposition VIII:2. Then

$$C_n \int_{|w|=1} u(Rw) d\sigma(w) - \log R \int_{|z| \leq R} dd^c u \wedge (dd^c \log^+|z|)^{n-1} =$$

$$= C_n \int_{|w|=1} u(w) d\sigma(w) - \int_{|z|<R} \log^+|z| dd^c u \wedge (dd^c \log^+|z|)^{n-1}$$

and the left hand side is, by Lemma VIII:1 not less than zero.

Hence,

$$0 \leq \int_{\mathbb{C}^n} \log^+|z| \, dd^c u \wedge (dd^c \log^+|z|)^{n-1} \leq C_n \int_{|u|=1} u(w) d\sigma(w).$$

So

$$C_n \int_{|w|=1} u(zw) d\sigma(w) - \log R \int_{|z| \leq R} dd^c u \wedge (dd^c \log^+|z|)^{n-1} =$$

$$= C_n \int_{|w|=1} u(Rw) d\sigma(w) - \log R) + \log R [\int_{\mathbb{C}^n} dd^c u \wedge (dd^c \log^+|z|)^{n-1} -$$

$$- \int_{|z| \leq R} dd^c u \wedge (dd^c \log^+|z|)^{n-1}] =$$

$$= C_n [\int_{|w|=1} u(Rw) d\sigma(w) - \log R] + \log R \int_{|z| \geq R} dd^c u \wedge (dd^c \log^+|z|)^{n-1}$$

and

$$\log R \int_{|z| \geq R} dd^c u \wedge (dd^c \log^+|z|)^{n-1} \to 0; \quad R \to +\infty$$

since

$$\int_{\mathbb{C}^n} \log^+|z| \wedge dd^c u \wedge (dd^c \log^+|z|)^{n-1} < +\infty$$

which proves the theorem.

Corollary VIII:1. If $\log^+|z| \leq u \in L$ then

$$\int \log^+|z| \wedge dd^c u \wedge (dd^c \log^+|z|)^{n-1} < +\infty.$$

Lemma VIII:2. If u_i is a decreasing sequence of functions in L such that $\lim u_i = u_0 \geq \log^+|z|$ then

$$\lim \gamma(u_j) = \gamma(u_0).$$

Proof. It is clear from the definition that $\lim \gamma(u_j) \geq \gamma(u_0)$. On the other hand, we have from Theorem VIII:2 that

$$\lim_{j \to +\infty} \gamma(u_j) = \int_{|w|=1} u_0 d\sigma(w) - \frac{1}{C_n} \lim_{j \to +\infty} \int_{\mathbb{C}^n} \log^+|z| \, dd^c u_j \wedge (dd^c \log^+|z|)^n \leq$$

$$\leq \int_{|w|=1} u_0 d\sigma(w) - \frac{1}{C_n} \lim_{j \to +\infty} \int_{|z|<R} \log^+|z| \, dd^c u_j \wedge (dd^c \log^+|z|)^{n-1} \leq$$

$$\leq \int_{|w|=1} u_0 d\sigma(w) - \frac{1}{C_n} \int_{|z|<R} \log^+|z| \, dd^c u_0 \wedge (dd^c \log^+|z|)^{n-1}$$

by Theorem V:2.

If we let $R \to +\infty$ and apply Theorem VIII:2 again we get the desired conclusion.

Definition. Let $u \in L$, we then define $\Gamma_r(u) = \sup_{|z|=r} u(z) - \log r$ and $\Gamma(u) = \lim_{r \to +\infty} \Gamma_r(u)$ (since $\Gamma_{e^r}(u)$ is convex and bounded at $+\infty$, $\Gamma_r(u)$ is decreasing in r).

It is clear that $\gamma(u) \leq \Gamma(u)$. If $E \subset B$, B the unit ball in $\mathbb{C}$, then $\gamma(V_E) = \Gamma(V_E)$. The following example shows that if $n=2$, we can have strict inequality.

Example 1. Fix $a>1$ and take $K=\{(z_1,z_2)\in\mathbb{C}^2;\ |z_1|^2+a|z_2|^2\leq 1\}$. Then $V_K^*(z) = \frac{1}{2}\log^+(|z_1|^2+a|z_2|^2)$ so

$$\gamma(V_K^*)= \frac{1}{2}\int_{|w|=1}\log(|w|^2+a|w_2|^2)\,d\sigma(w) < \frac{1}{2}\log a = \Gamma(V_K^*).$$

Proposition VIII:3. Let $\Theta_j=(dd^c u)^{n-j}\wedge(dd^c\log^+|z|)^j$, $1\leq j\leq n$, $\log^+|z|\leq u\in L$.

Then $\int u\Theta_j \leq \int \log^+|z|(dd^c u)^n + C_n j\,\Gamma(u).$

Proof. If we first take $j=1$, Proposition VIII:2 gives

$$\int u\,dd^c\log^+|z|\wedge(dd^c u)^{n-1}=\int \log^+|z|(dd^c u)^n +$$

$$+ \int_{|z|=R} u\,d^c\log^+|z|\wedge(dd^c u)^{n-1}-\log R\int_{|z|=R} d^c u\wedge(dd^c u)^{n-1} \leq C_n\Gamma_R(u) +$$

$$+ \log R\int_{|z|\geq R}(dd^c u)^n \to C_n\Gamma(u),\ R\to+\infty\quad\text{since}\quad \log R\int_{|z|>R}(dd^c u)^n\to 0,\ R\to+\infty.$$

Assuming the proposition for j, we want to prove it for $j+1$.

$$\int u\Theta_{i+1}=\int u(dd^c u)^{n-j-1}\wedge(dd^c\log^+|z|)^{j+1} =$$

$$= \int u(dd^c\log^+|z|)\wedge(dd^c u)^{n-j-1}\wedge(dd^c\log^+|z|)^j \leq (\text{Prop. VIII:2})$$

$$\leq \int \log^+|z|(dd^c u)^{n-j}\wedge(dd^c\log^+|z|)^j + C_n\Gamma(u) \leq$$

$$\leq \int u(dd^c u)^{n-j} \wedge (dd^c \log^+|z|)^j + C_n \Gamma(u) \leq \text{ (assumption)}$$

$$\leq \int u(dd^c \log^+|z|)^n + C_n j\Gamma(u) + C_n \Gamma(u) = \int u(dd^c \log^+|z|)^n +$$

$$+ C_n(j+1)\Gamma(u).$$

Corollary VIII:1. If $\log^+|z| \in u \in L$ then

$$\int_{|w|=1} u(w)\,d\sigma \leq \frac{1}{C_n} \int \log^+|z|(dd^c u)^n + \gamma(u) + (n-1)\Gamma(u).$$

In particular if $E \subset B$, the unit ball, then

$$\int_{|w|=1} V_E^*(w)\,d\sigma(w) \leq \gamma(V_E^*) + (n-1)\Gamma(V_E^*).$$

Proof. Take $j=n-1$ in Proposition VIII:3. Then, by Theorem VIII:2, we get

$$\int_{|w|=1} u(w)\,d\sigma(w) - \gamma(u) = \frac{1}{C_n} \int_{\mathbb{C}^n} \log^+|z|\,dd^c u \ (dd^c \log^+|z|)^{n-1} \leq$$

$$\leq \frac{1}{C_n} \int u\,dd^c u \wedge (dd^c \log^+|z|)^{n-1} \leq \frac{1}{C_n} \int \log^+|z|(dd^c u)^n +$$

$$+ C_n(n-1)\Gamma(u)$$

which proves the first statement. The second statement follows from the fact that if $E \subset B$ then $\operatorname{supp}(dd^c V_E^*)^n \subset \bar{B}$.

Remarks and references

A proof of Proposition VIII:1 is in **J. Siciak,** Extremal plurisubharmonic functions in $\mathbb{C}^n$, Proceedings of the first Finnish-Polish Summerschool in Complex Analysis in Podlesice, 1977, pg. 123-124.

Theorem VIII:1 is due to **N. Levenberg,** Monge-Ampère Measures Associated to Extremal Plurisubharmonic Functions in $\mathbb{C}^n$. Trans. Am. Math. Soc. 289 (1985), 333-343.

Lemma VIII:1 is proved by **B.A. Taylor,** An estimate for an extremal plurisubharmonic function. Séminare d'Analyse P. Lelong, Dolbeault-H. Skoda, 1981/1983. Springer Lecture Notes in Mathematics 1028. This paper also contains a somewhat weaker version of Corollary VIII:1.

In **S. Kołodziej,** The logarithmic capacity in $\mathbb{C}^n$ (To appear in Ann. Pol. Math.), it was proved that $C(E) = e^{-\Gamma(V_E)}$ is a capacity in Choquets sense. That $e^{-\gamma(V_E)}$ is a capacity was proved by the same author in: Capacities associated to the Siciak extremal function, Manuscript. Cracow. 1986.

The relationship between γ and Γ has also been studied by **J. Siciak,** On logarithmic capacities and pluripolar sets in $\mathbb{C}^n$. Manuscript, October 1986.

Using Corollary 6.7 in **E. Bedford** and **B.A. Taylor,** Plurisubharmonic functions with logarithmic singularities, Manuscript 1987, one can prove that $e^{-\Gamma(V_E)}$ is an outer regular capacity.

V.P. Zaharjuta has studied capacities and extremal plurisubharmonic functions in connection with transfinite diameter

and the Bernstein-Walsh theorem: Transfinite diameter Čebychêv
constants and capacity for compact in $\mathbb{C}^n$. Math. USSR Sbornik,
Vol. 25 (1975), No. 3.

Extremal plurisubharmonic functions, orthogonal polynomials
and the Bernstein-Walsh theorem for analytic functions of
several complex variables. Ann. Polon. Math. 33 (1976).

For results and more references, see:

Nguyen Thanh Van and **Ahmed Zeriahi**, Familles de polynômes
presque partout bornées. Bull. Sc. Math. 2c Série 107 (1983).

W. Plesniak and **W. Pawłucki**, Markov´s inequality and C^∞
functions on sets with polynomial cusps. Math. Ann. 275 (1986),
467-480.

A. Sadullaev, Plurisubharmonic measures and capacities on
complex manifolds. Russian Math. Surveys 36 (1981).

IX Gamma Capacity

Definition (the Choquet Integral). Assume that f is a non-negative function and c a capacity. Then $\int fd_c$ is defined by

$$\int fd_c = \int_0^\infty c(\{x;\ f(x)>s\})ds.$$

Definition. Let p be a precapacity. A non-negative function f is said to be p-capacitable if

$$\int fd_p = \sup \int gd_p;\ g\leq f,\quad g\quad \text{upper semi-continuous}.$$

Lemma IX:1. Assume that E_ν, $\nu\in N$ is an increasing sequence of p-capacitable sets. Then $\bigcup_{\nu=1}^\infty E_\nu$ is p-capacitable.

Proof. We have $p(\bigcup_{\nu=1}^\infty E_\nu) = \sup_{\nu\in N} p(E_\nu)$. Let $\varepsilon>0$ be given. Choose ν so that

$$p(\bigcup_{\nu=1}^\infty E_\nu) < p(E_\nu) + \varepsilon/2$$

and a compact subset k of E_ν such that

$$p(E_\nu) < p(k) + \varepsilon/2.$$

Then k is compact in $\bigcup_{\nu=1}^\infty E_\nu$ and

$$p(\bigcup_{\nu=1}^\infty E_\nu) < p(k) + \varepsilon$$

so $\bigcup_{\nu=1}^\infty E_\nu$ is p-capacitable.

Theorem IX:1. If f is p-capacitable then $\{x; f(x)>s\}$ is p-capacitable.

Proof. Assume that f is p-capacitable. Then there is an increasing sequence $\{f_n\}_{n=1}^{\infty}$ of upper semi-continuous functions which are smaller or equal to f with

$$\lim_{n\to+\infty} \int f_n d_p = \int f d_p.$$

It is no restriction to assume that every f_n has compact support.

It is easy to see that

$$p(\{x; f(x)>s\}) = p(\{x; \lim_{n\to+\infty} f_n(x)>s\}) \quad \forall s\geq 0.$$

Put $E_{m,n}=\{x; f_n(x)\geq s+\frac{1}{m}\}$. Every $E_{m,n}$ is a compact subset of $\{x; f(x)>s\}$ and

$$\{x; \lim_{n\to+\infty} f_n(x)>s\} = \bigcup_{m=1}^{\infty} \bigcup_{n=1}^{\infty} E_{m,n}$$

so it follows from Lemma IX:1 that $\{x; \lim_{n\to+\infty} f_n(x)>s\}$ is p-capacitable and $\{x; f(x)>s\}$ has to be p-capacitable.

Theorem IX:2. Assume that c is a strongly subadditive capacity. Denote by χ_A the characteristic function of A. Put

$$\tilde{\beta}_1(f)=\inf(\sum_{i=1}^{n} \alpha_i c(A_i); \sum_{i=1}^{n} \alpha_i \chi_{A_i} \geq f).$$

Then

$$\tilde{\beta}_1(f) = \int fd_c.$$

Corollary IX:1. Assume that c is a capacity as in Theorem IX:2. Then the Choquet integral is subadditive (and therefore a seminorm on the non-negative functions).

Corollary IX:2. Assume that c is a capacity as in Theorem IX:2. Then

$$\int fd_c = \inf \int gd_c; \quad f \leq g, \quad g \text{ is lower semi-continuous.}$$

<u>Swarms. Product capacities.</u>

Definition. A class of functions $(L_E)_{E \subset V}$

$$L_E: \ U \to [0,+\infty]$$

is called a swarm if

α) for every fixed $x \in U$, $E \to L_E(x)$ is a capacity on V,

β) for every compact subset K of V, $L_K(x)$ is a bounded, upper semi-continuous function with compact support.

Example. Choose $U=V=\mathbb{C}^n$ and denote by χ_E the characteristic function of E. Then $(\chi_E)_{E \subset \mathbb{C}^n}$ is a swarm.

Theorem IX:3. Assume that c is a capacity on U and $(L_E)_{E \subset V}$ a swarm. Then $C(E) = \int L_E d_c$ is a capacity on V. Furthermore, if $L_E(x)$ is subadditive for all x in U and if c is strongly subadditive then C is subadditive.

Proof. i) clear.

ii) Let E_ν, $\nu \in \mathbb{N}$, be a non-decreasing sequence of subsets of V. Then

$$\sup_{\nu \in \mathbb{N}} C(E_\nu) = \lim_{\nu \to +\infty} \int L_{E_\nu} \, dc = \lim_{\nu \to +\infty} \int_0^\infty c(\{x \in U; \ L_{E_\nu}(x) > s\}) ds =$$

$$= \int_0^\infty \lim_{\nu \to +\infty} c(\{x \in U; \ L_{E_\nu} > s\}) ds = \int_0^\infty c(\{x \in U; \ L_{\bigcup_{\nu=1}^\infty E_\nu}(x) > s\}) ds =$$

$$= C(\bigcup_{\nu=1}^\infty E_\nu).$$

iii) Let K_ν, $\nu \in \mathbb{N}$, be a decreasing sequence of compact subsets of V. Then

$$\inf_{\nu \in \mathbb{N}} C(K_\nu) = \lim_{\nu \to +\infty} \int_0^\infty c(\{x \in U \ ; \ L_{K_\nu}(x) > s\}) ds =$$

$$= \lim_{\nu \to +\infty} \int_0^\infty c(\{x \in U; \ L_{K_\nu}(x) \geq s\}) ds = \int_0^\infty \lim_{\nu \to +\infty} c(\{x \in U; \ L_{K_\nu}(x) \geq s\}) ds =$$

$$= \int_0^\infty c(\{x \in U; \ L_{\bigcap_{\nu=1}^\infty K_\nu}(x) \geq s\}) ds = C(\bigcap_{\nu=1}^\infty K_\nu).$$

The last statement in the theorem follows from Corollary IX:1.

Theorem IX:4. Let $(L_E)_{E \subset V}$ be a swarm. Then L_E is a universally capacitable function for every universally capacitable set $E \in P(V)$.

Proof. Since, by Theorem IX:3, $C(E)=\int L_E d_c$ is a capacity for every capacity c, we have for any universally capacitable set E

$$\int L_E d_c = C(E) = \sup_{\substack{K \subset E \\ K \text{ compact}}} C(K) = \sup_{\substack{K \subset E \\ K \text{ compact}}} \int L_K d_c .$$

Since L_K is upper semi-continuous and less or equal to L_E, L_E is c-capacitable. But c was an arbitrarily chosen capacity so it follows that L_E is universally capacitable.

Theorem IX:5. Assume that c is a capacity on V. Then $L_E(x)=c(\{y \in V;\ (x,y) \in E\})$, $E \subset U \times V$ is a swarm. Furthermore, if c is subadditive then $L_E(x)$ is subadditive for every fixed x in U.

Proof. α) clear.

β) Let a compact subset K of $U \times V$ be given. It is clear that L_K has compact support so it remains to prove that L_K is upper semicontinuous. Given $\alpha > 0$ and $x_0 \in \overline{\{x \in U;\ L_K(x) \geq \alpha\}}$. We have to prove that $L_K(x_0) \geq \alpha$. Choose x_n with $L_K(x_n) \geq \alpha$ such that $x_n \to x_0$, $n \to +\infty$. Put

$$D_0 = \{y \in V;\ (x_0,y) \in K\}$$

and

$$D_n = \{y \in V;\ (x_n,y) \in K\} .$$

It is easily verified that

$$D_0 \supset \bigcap_{i=1}^{\infty} \overline{\left(\bigcup_{j=i}^{\infty} D_j \right)}.$$

Since c is a capacity we have

$$L_K(x_0)=c(D_0) \geq c\left(\bigcap_{i=1}^{\infty} \overline{\left(\bigcup_{j=i}^{\infty} D_j \right)} \right)=$$

$$= \lim_{i \to +\infty} c\left(\overline{\bigcup_{j=i}^{\infty} D_j} \right) \geq \overline{\lim_{j \to +\infty}} \, c(D_j) = \overline{\lim_{j \to +\infty}} \, L_K(x_j) \geq \alpha.$$

The last statement in the theorem is obvious and the proof is complete.

Definition (Product Capacity). Let c and d be capacities on U and V respectively. The product capacity $c \times d$ on $U \times V$ is defined by

$$c \times d(E) = \int L_E d_c$$

where

$$L_E(x)=d(\{y \in V; \; (x,y) \in E\}).$$

By Theorems IX:5 and IX:3, $c \times d$ is a capacity. Furthermore, if c is strongly subadditive and d is subadditive, it follows from Theorem IX:2 that $c \times d$ is subadditive.

Example I. Let $U=V=\mathbb{C}$ and denote by c the Newtonian capacity. Consider in $\mathbb{C}^2$ the set $E=\{(z_1,z_2); \; |z_1|+|z_2|=1, \; \mathrm{Im} \, z_1=0\}$. It is easily seen that $c \times c(E)>0$. But if we interchange the variables z_1 and z_2, i.e. consider the set $E'=\{(z_1,z_2); \; |z_1|+|z_2|=1; \; \mathrm{Im} \, z_2=0\}$ it is clear that $L_{E'}(z_1)=0$, $\forall z_1 \in \mathbb{C}$, so $c \times c(E')=0$.

Let now U be an open subset of $\mathbb{C}$ and c a capacity on U. We construct c_n on U^n by induction

$$c_1 = c \qquad \text{on} \quad U$$
$$\vdots$$
$$c_n = c \times c_{n-1} \qquad \text{on} \quad U^n.$$

It is clear that c_n is a capacity on U^n, if c is strongly subadditive, then c_n is subadditive. Put for $E \subset U^n$, $x \in U^{n-p}$

$$L_E^{n-p}(x) = c_p(\{y \in U^p;\ (x,y) \in E\}.$$

By Theorem IX:5, $(L_E^{n-p})_{E \subset U^n}$ is a swarm.

Remark. It follows from Theorem IX:4 and Theorem IX:1 that $\{x \in U^{n-p};\ L_E^{n-p}(x) > s\}$ is universally capacitable for every $s \geq 0$ and every universally capacitable set E.

Let c be a capacity on an open subset U of $\mathbb{C}$. We define a precapacity P_n on U^n by induction

$$P_1 = c \qquad \text{on} \quad U^1$$
$$\vdots$$
$$P_n(E) = c(\{x \in U;\ P_{n-1}(\{y \in U^{n-1};\ (x,y) \in E\}) > 0\}),\ E \subset U^n.$$

Theorem IX:6. 1) $P_n(E) = 0 \iff c_n(E) = 0$,

2) $P_n(E) = c(\{x \in U;\ L_E^1(x) > 0\})$,

3) P_n is a precapacity on U^n,

4) every universally capacitable is P_n-capacitable,

5) if c is subadditive, the P_n is subadditive.

Proof. 1), 2) induction. n=1 clear. Assume that 1) and 2) hold for n-1. Prove 1) and 2) for n.

$$P_n(E) = c(\{x \in U; \ P_{n-1}(\{y \in U^{n-1}; \ (x,y) \in E\}) > 0\} =$$

$$= c(\{x \in U; \ c_{n-1}(\{y \in U^{n-1}; \ (x,y) \in E\}) > 0\}) =$$

$$= c(\{x \in U \ ; \ L_E^1(x) > 0\}).$$

Thus

$$P_n(E) = c(\{x \in U; \ L_E^1(x) > 0\})$$

and it is clear that $c\{x \in U; \ L_E^1(x) > 0\} = 0$ if and only if $c_n(E) = \int L_E^1 d_c = 0$. This proves 1) and 2).

3) i)-iii) are clear since

$$P_n(E) = c(\{x \in U; \ L_E^1(x) > 0\})$$

where c is a capacity and $(L_E^1)_{E \in P(U^n)}$ is a swarm.

4) Assume that E is universally capacitable. By Theorem IX:4, L_E^1 is universally capacitable and

$$\int L_E^1 d_c = \int \lim_{\nu \to +\infty} L_{K_\nu}^1 \, d_c$$

where K_ν, $\nu \in \mathbb{N}$, is an increasing sequence of compact subsets of E. Hence

$$c(\{x \in U; \ L_E^1(x) > s\}) = c(\{x \in U; \ \lim_{\nu \to +\infty} L_{K_\nu}^1(x) > s\}) =$$

$$= \lim_{\nu \to +\infty} c\{x \in U; \ L_{K_\nu}(x) > s\})$$

for all $s \geq 0$, so $P_n(E) = \lim_{\nu \to +\infty} P_n(K_\nu) = \sup\{P(K); K \text{ compact}, K \subset E\}$ which means that E is P_n-capacitable.

5) Induction. $n=1$ clear. Assume that P_{n-1} is subadditive. Then $P_n(E_1 \cup E_2) = c(\{x \in U; \, P_{n-1}(\{y \in U^{n-1}; \, (x,y) \in E_1 \cup E_2\}) > 0) =$

$= c(\{x \in U; \, P_{n-1}(\{y \in U^{n-1}; (x,y) \in E_1\} \cup \{y \in U^{n-1}; (x,y) \in E_2\} > 0)\} \leq$

$\leq c(\{x \in U; \, P_{n-1}(\{y \in U^{n-1}; \, (x,y) \in E_1\}) > 0\} \cup$

$\cup \{x \in U; \, P_{n-1}(\{y \in U^{n-1}; \, (x,y) \in E_2\}) > 0\}) \leq P_n(E_1) + P_n(E_2)$ and it follows that P_n is subadditive.

Corollary IX:3. Assume that c is a capacity on U. If $c(E_\nu) = 0$, $\nu = 1,2 \Rightarrow c(\bigcup_{\nu=1}^{2} E_\nu) = 0$, then c_n and P_n has the same property.

Proof. As the proof of Theorem IX:6, 2.

Theorem IX:7. Let c be a (pre)capacity on V and $(\alpha_i)_{i \in I}$ a complete normal family of continuous functions, $\alpha_i : U \to V$. Then

$$C(E) = \sup_{i \in I} c(\alpha_i(E))$$

is a (pre)capacity on U. If c is subadditive, then C is subadditive.

Proof. i), ii) clear.

iii) Let E_ν, $\nu \in \mathbb{N}$, be an increasing sequence of subsets of U. Given $\varepsilon > 0$. Choose α_{i_ε} such that $C(E) < c(\alpha_{i_\varepsilon}(E)) + \varepsilon/2$ and then E_ν such that $c(\alpha_{i_\varepsilon}(E)) < c(\alpha_{i_\varepsilon}(E_\nu)) + \varepsilon/2$. Then

$$C(E) < c(\alpha_{i_\varepsilon}(E_\nu)) + \varepsilon \leq C(E_\nu) + \varepsilon$$

and iii) is proved.

Since $\alpha_i(E_1 \cup E_2) = \alpha_i(E_1) \cup \alpha_i(E_2)$; $i \in I$, we have, for c subadditive

$$C(E_1 \cup E_2) = \sup_{i \in I} c(\alpha_i(E_1 \cup E_2)) = \sup_{i \in I} c(\alpha_i(E_1) \cup \alpha_i(E_2)) \leq$$

$$\leq C(E_1) + C(E_2)$$

which proves that C is subadditive.

Assume now that c is a capacity. Let K_ν, $\nu \in \mathbb{N}$, be a decreasing sequence of compact subsets of U. We have to prove that

$$\inf_{\nu \in \mathbb{N}} C(K_\nu) = C(\bigcap_{\nu=1}^{\infty} K_\nu).$$

Choose α_n so that

$$C(K_n) < c(\alpha_n(K_n)) + \frac{1}{n} .$$

Since $(\alpha_i)_{i \in I}$ is normal, and complete, we can assume that $\alpha_n \to \alpha_0$, uniformly on compact subsets of U. We claim that

$$\alpha_0(\bigcap_{n=1}^{\infty} K_n) \supset \bigcap_{i=1}^{\infty} (\overline{\bigcup_{n=i}^{\infty} \alpha_n(K_n)}).$$

Given

$$z \in \bigcap_{i=1}^{\infty} (\overline{\bigcup_{n=i}^{\infty} \alpha_n(K_n)}).$$

Then

$$z \in \overline{\bigcup_{n=i}^{\infty} \alpha_n(K_n)} \qquad \forall i \in N.$$

Choose $z_\ell \in \alpha_{n(\ell)}(K_{n(\ell)})$ where $n(\ell) > \ell$ such that $z_\ell \to z$, $\ell \to +\infty$. Then $z_\ell = \alpha_{n(\ell)}(x_\ell)$ where $x_\ell \in K_{n(\ell)}$ and we can assume that $x_\ell \to x \in \bigcap_{\nu=1}^{\infty} K_\nu$, $\ell \to +\infty$. Now

$$|z - \alpha_0(x)| \leq |z - z_\ell| + |z_\ell - \alpha_0(x)| \leq$$

$$\leq |z - z_\ell| + |\alpha_{n(\ell)}(x_\ell) - \alpha_0(x)| \leq$$

$$\leq |z - z_\ell| + |\alpha_{n(\ell)}(x_\ell) - \alpha_0(x_\ell)| + |\alpha_0(x_\ell) - \alpha_0(x)| \to 0, \quad \ell \to +\infty,$$

so $z = \alpha_0(x)$ and since z was an arbitrary element in $\bigcap_{i=1}^{\infty} (\overline{\bigcup_{n=i}^{\infty} \alpha_n(K_n)})$ we have proved that

$$\alpha_0(\bigcap_{n=1}^{\infty} K_n) \supset \bigcap_{i=1}^{\infty} (\overline{\bigcup_{n=i}^{\infty} \alpha_n(K_n)}).$$

To finish the proof, we can now argue as in the end of the proof of Theorem IX:5.

Ronkin´s gamma-capacity and Favorov's capacity

Denote by $\underline{\mathrm{cap}}_2$ and $\overline{\mathrm{cap}}_2$ the interior and exterior logarithmic capacity on $\mathbb{C}$, respectively. γ_n is then defined as follows

$$\gamma_1(E) = \underline{cap}_2(E) \qquad\qquad E \subset \mathbb{C}$$

$$\vdots$$

$$\gamma_n(E) = \underline{cap}_2(\{z_1 \in \mathbb{C};\ \gamma_{n-1}(\{z \in \mathbb{C}^{n-1};\ (z_1,z) \in E\}) > 0\}),\ E \subset \mathbb{C}^n$$

Ronkin's gamma-capacity is then by definition

$$\Gamma_n(E) = \sup\{\gamma_n(\alpha(E));\ \alpha \quad \text{complex unitary transformation}\}.$$

Proposition IX:1. $\gamma_n(E) = P_n(E)$ for all universally capacitable sets E, where P_n is formed with respect to $\overline{cap}_2$.

Proof. Induction. Clearly Proposition IX:1 is true for $n=1$.

Assume it is true for $n-1$. Prove it for n.

Assume that E is universally capacitable.

$$\gamma_n(E) = \underline{cap}_2(\{z_1;\ \gamma_{n-1}(\{z \in \mathbb{C}^{n-1};\ (z_1,z) \in E\}) > 0\}).$$

It is clear that $\{z \in \mathbb{C}^{n-1};\ (z_1,z) \in E\}$ is universally capacitable. Hence

$$\gamma_{n-1}(\{z \in \mathbb{C}^{n-1};\ (z_1,z) \in E\}) = P_{n-1}(\{z \in \mathbb{C}^{n-1};\ (z_1,z) \in E\})$$

for all $z_1 \in \mathbb{C}$. Then

$$\gamma_n(E) = \underline{cap}_2(\{z_1 \in \mathbb{C};\ P_{n-1}(\{z \in \mathbb{C}^{n-1};\ (z_1,z) \in E\}) > 0\}) =$$

$$= (\text{Theorem IX:6, 1)}) = \underline{cap}_2(\{z_1 \in \mathbb{C};\ C_{n-1}(\{z \in \mathbb{C}^{n-1};\ (z_1,z) \in E\}) > 0\}) =$$

$$= (\text{by definition}) = \underline{cap}_2(\{z_1 \in \mathbb{C};\ L_E^1(z_1) > 0\}) =$$

$$= (\text{Remark p. 66}) = \overline{cap}_2(\{z_1 \in \mathbb{C};\ L_E^1(z_1) > 0\}) = P_n(E)$$

and the proposition is proved.

Corollary IX:3. If E is universally capacitable then

$$\gamma_n(E)=\overline{\mathrm{cap}}_2(\{z_1\in\mathbb{C};\ \gamma_{n-1}(\{z\in\mathbb{C}^{n-1};\ (z_1,z)\in E\})>0\}).$$

Corollary IX:4. By Corollary IX:3, $\Gamma_n(E_1)=\Gamma_n(E_2)=0$ implies that

$$\Gamma_n(E_1\cup E_2)=0.$$

Proposition IX:2. Every universally capacitable set is Γ_n-capacitable.

Proof. Assume that E is universally capacitable and let $\varepsilon>0$ be given. Choose a complex unitary transformation α such that

$$\Gamma_n(E) < \gamma_n(\alpha(E))+\varepsilon/2.$$

$\alpha(E)$ is universally capacitable so by Theorem IX:6 there is a compact subset K of $\alpha(E)$ such that

$$\gamma_n(\alpha(E)) < \gamma_n(K)+\varepsilon/2.$$

Thus $\Gamma_n(E)<\Gamma_n(\alpha^{-1}(K))+\varepsilon$ and since $\alpha^{-1}(K)$ is a compact subset of E, the proof is complete.

Favorov´s capacity is by definition

$$\Gamma_n^F(E) = \sup\{c_n(\alpha(E));\ \alpha \text{ complex unitary tranformation}\},$$

where c is the logarithmic capacity on $\mathbb{C}$.

Corollary IX:4. The analogy of Corollary IX:5 holds true for Γ_n^F.

Proof. Follows from Theorem IX:6 and Corollary IX:4.

Proposition IX:2. If E is $\mathbb{C}^n$-polar then $\Gamma_n^F(E)=0$.

Proof. The property being $\mathbb{C}^n$-polar is invariant under complex unitary transformations. It is therefore enough to prove that $\gamma_n(E)=0$. Assume the proposition to be true for $n-1$. Then by Theorem IX:6 we have

$$\gamma_n(E)=\mathrm{cap}_2\{z_1\in\mathbb{C};\ \gamma_{n-1}(\{(z_1,\ldots,z_n)\in E\})>0\} =$$

$$= \mathrm{cap}_2\{z_1\in\mathbb{C};\ \varphi(z_1,\ldots,z_n)\equiv-\infty,)\ (z_2,\ldots,z_n)\in\mathbb{C}^{n-1}\}$$

where $\varphi\not\equiv-\infty$ is a plurisubharmonic function such that $E\subset\{\varphi=-\infty\}$. But a subharmonic function $(\not\equiv-\infty)$ on $\mathbb{C}$ is $>-\infty$ outside a set of $\mathrm{cap}_2=0$ which proves the proposition.

The following example shows that there are non-$\mathbb{C}^n$-polar sets of zero gamma-capacity.

Example 2. Put $H=\{z\in\mathbb{C}^2;\ \mathrm{Im}\ z_1=\mathrm{Re}(z_1+z_2)=0\}$. It is clear that $\Gamma_n(H)>0$. Denote by g the biholomorphic map

$$g(z_1,z_2)=(z_1-z_2^2,z_2).$$

Then

$$g(H)=\{\omega\in\mathbb{C}^2;\ \mathrm{Im}(\omega_1+\omega_2^2) = \mathrm{Re}(\omega_1+\omega_2+\omega_2^2)=0\}.$$

Any complex line cuts $g(H)$ in at most four points so $\Gamma_2(g(H))=0$.

Now, by Proposition IX:2, H is not $\mathbb{C}^2$-polar and since g is biholomorphic, $g(H)$ cannot be $\mathbb{C}^2$-polar. Hence, there are non-$\mathbb{C}^2$-polar subsets of $\mathbb{C}^2$ with vanishing Γ_2-capacity.

To get a capacity with null sets invariant under holomorphic mappings we proceed as follows.

Let B denote an open and bounded subset of $\mathbb{C}$. Denote by A_n the holomorphic mappings f, $f: B^n \to B^n$.

Definition. $h_n(E) = \sup_{f \in A_n} \Gamma_n^F(f(E))$, $E \subset B^n$.

Theorem IX:8.

i) $E \subset B^n ==> h_n(E) \geq h_n(f(E))$, $f \in A_n$,

ii) h_n vanishes on $\mathbb{C}^n$-polar subsets of B^n.

Proof. i) follows from the definition of h_n. Observe that this means that h_n is invariant under biholomorphic mappings of B^n onto itself.

ii) Assume that N is a $\mathbb{C}^n$-polar subset of B^n. It follows from Proposition IX:2 that $\Gamma_n^F(N)=0$. Let now $f \in A_n$ be given. We have to prove that

$$\Gamma_n^F(f(N))=0.$$

Denote by $\tau(f)$ the Jacobian of f. It is clear that

$$\Gamma_n^F(f(N \cap \{\tau(f) \neq 0\})) = 0$$

- 96 -

so by Corollary IX:5 it remains to prove that

$\Gamma_n(f(N\cap\{\tau(f)=0\}))=0$. This follows from Corollary IX:6 below.

Definition. A subset E of $\mathbb{C}^n$ is called a (proper) locally analytic set if for every ω in E there is a neighborhood O_ω of ω such that $E\cap O_\omega$ is a (proper) analytic set in O_ω.

Theorem IX:9. Let U be an open subset of $\mathbb{C}^n$ and $F=(f_1,\ldots,f_n)$ a holomorphic map $F:U\to\mathbb{C}^n$. Then $F(\{\tau(F)=0\})$ is contained in a denumerable union of proper locally analytic sets.

Proof. Put $J=\{z\in U;\ \tau(F)=0\}$. We claim that

$$P \quad \text{analytic of} \quad \dim\leq m \implies F(P\cap J)$$

is contained in a denumerable union of proper locally analytic sets.

This is clearly true for $m=0$, for an analytic set of zero dimension is denumerable. Now, if the statement is true for $m-1$ we have to prove it for P with $\dim P=m$. We can assume that $P\cap J$ is connected.

Choose $\left(\dfrac{\partial f_{ip}}{\partial z_{iq}}\right)_{\substack{p=1,\ldots,s \\ q=1,\ldots,s}}$ with

$$G=\det\left(\frac{\partial f_{ip}}{\partial z_{iq}}\right)\neq 0 \quad \text{where} \quad s=\max\ \mathrm{rank}\left(\frac{\partial f_i}{\partial z_j}\bigg|_{P\cap J}\right)_{i,j=1,\ldots,n}.$$

Put $Q=(J\cap P)_{reg}\cap\{G\neq 0\}$. By the remark in Remmert (see the notes below), $F(Q)$ is contained in a denumerable union of proper locally analytic sets. $Q^1=(J\cap P)_{reg}\cap\{G=0\}$ is of dimension $\leq m-1$ so, by assumption, $F(Q')$ is contained in a denumerable union of proper locally analytic sets.

Furthermore, since $\dim(J\cap P)_{sing}\leq m-1$, the same is true for $F(J\cap P)_{sing})$. Since

$$F(J\cap P)\subset F((J\cap P)_{sing})\cup F(Q)\cup F(Q')$$

the statement is proved.

To prove the theorem, it is enough to choose $P=U$.

Corollary IX:6. If N is $\mathbb{C}^2$-polar in U, where U is open in $\mathbb{C}^n$ and if $F: U\to\mathbb{C}^n$ is an analytic map, then $F(N)$ is $\mathbb{C}^n$-polar in $\mathbb{C}^n$.

Proof. It is clear that $F(N\cap\{\tau(F)\neq 0\})$ is $\mathbb{C}^n$-polar. From Theorem IX:9 it follows that $F(N\cap\{\tau(F)=0\})$ is $\mathbb{C}^n$-polar, since proper locally analytic sets are $\mathbb{C}^n$-polar.

<u>Notes and references</u>

Theorem IX:2 is due to **F. Topsøe,** On construction of measures. Københavns Universitet, Mat. Inst. Preprint series 1974:27.

Corollary IX:1 is due to **G. Choquet,** Theory of capacities. Ann. Inst. Fourier 5 (1953-54).

The notation of swarm is closely related to that of "noyau capacitaire regulier" as defined in **C. Dellacherie,** Ensembles analytiques. Capacités. Mesures de Hausdorff. Springer LNM. 295 (1972).

Example 1 is due to **V. Šeinow** (see Ronkins book below).

Example 2 is due to **C.O. Kiselman.** Manuscript. Uppsala 1973.

The gamma capacity was introduced in **L.I. Ronkin,** Introduction to the theory of entire functions of several variables. Amer. Math. Soc. Providence. R.I. 1974, and the modified gamma capacity is in **S.Ju. Favorov,** On capacity characterizations of sets in $\mathbb{C}^n$. Charkov 1974 (Russian). The remark by Remmert, used in the proof of Theorem IX:9 is in **R. Remmert,** Holomorphe und meromorphe Abbildungen komplexer Räume. Math. Ann. 133 (1957).

The set functions γ_n and Γ_n has been used in connection with removable singularity sets; cf. **U. Cegrell,** Removable singularity sets for analytic functions having modulus with bounded Laplace mass. Proc. Amer. Math. Soc. Vol. 88 (1983).

P. Järvi, Removable singularities for H^p-functions. Proc. Amer. Math. Soc. Vol. 86 (1982).

J. Riihentaus, An extension theorem for meromorphic functions of several variables. Ann. Acad. Sc. Fenn. Sér. AI. Vol. 4 (1978/79).

Some of the material of this section has been published in seminaire Pierre Lelong - Henri Skoda (Analyse) 1978/79. Springer LNM 822. 1980.

X Capacities on the Boundary

Let U be an open, bounded and connected subset of $\mathbb{C}^n$ containing zero. Then H(U) (H$^\infty$(U)) is the class of (bounded) analytic functions on U and A(U) consists of the functions in H(U) that extends continuously to $\overline{U}$. If μ is a positive measure on ∂U we define H^p(μ,∂U) (1$\leq$p<+∞) to be the closure of A(U) in L^p(μ,∂U).

For z$\in\overline{U}$ we define the classes of probability measures

$$M_z = \{\mu \geq 0; \ f(z) = \int f d\mu, \ \forall f \in A(U), \ \text{supp}\mu \subset \partial U\}$$

and

$$N_z = \{\mu \geq 0; \ \varphi(z) \leq \int \varphi d\mu, \ \forall \varphi \in \text{PSH}(U) \cap C(\overline{U}), \ \text{supp}\mu \subset \partial U\}$$

(PSH and C are the plurisubharmonic functions and the continuous functions respectively.)

We write B for the unit ball in $\mathbb{C}^n$ and σ is the normalized Lebesgue measure on ∂B. Then f$\in$H^p(σ,∂B) if and only if

$$\sup_{0<r<1} \int |C[rz,\xi]f(\xi)d\sigma(\xi)|^p d\sigma(z) < +\infty$$

where $C[z,\xi] = (1-\langle z,\xi\rangle)^{-n}$ is the Cauchy kernel. Hence, each f$\in$H^1(σ,∂B) extends to H(B).

The space LH^1 consists of the functions f in H(U)

such that $|f|$ has a plurisubharmonic majorant. A norm on $LH^1(U)$ is

$$\|f\|_{LH^1} = \inf\{h(0); \ h \in PH(U), \ |f| \leq h\}$$

where PH denotes the plurisubharmonic functions. This norm is called the Lumer norm.

For each point $z \in \overline{U}$ we define the capacities

$$Q_z(E) = \sup_{\mu \in M_z} \mu(E)$$

$$R_z(E) = \sup_{\mu \in N_z} \mu(E).$$

Observe that $R \leq Q$ and that R can vanish at sets of positive Q-capacity. See example below. (We write $R=R_0$ and $Q=Q_0$.)

We now wish to study Q and therefore extend it to a functional. Put $P = \mathrm{Re}\ A(U)$ and let $\overline{P}$ be the closure of P in $L^1_{loc}(U)$.

Let φ be a real-valued function on ∂U and put for $z \in \overline{U}$

$$\hat{\varphi}(z) = \sup_{\mu \in M_z} \int^* \varphi d\mu.$$

Theorem X:1. If φ is upper semi-continuous on ∂U, then

$$\hat{\varphi}(z) = \inf\{h(z); \ h \in \overline{P}, \ h \geq \varphi \text{ on } \partial U\}, \ z \in U.$$

Proof. The proof of Lemma III:3 applies.

Lemma X:1. If $0 \leq \varphi$ is upper semi-continuous, then $-\hat{\varphi}$ is continuous and plurisubharmonic on U.

Proof. By Theorem X:1

$$-\hat{\varphi}(z)=\sup\{-h(z); \ h \in P, \ h \geq \varphi\}=\sup\{h(z); \ h \in P, \ h \leq -\varphi\},$$

so it follows $(-\hat{\varphi})*(z) = \overline{\lim_{z' \to z}} -\hat{\varphi}(z')$ is plurisubharmonic; it remains to prove that $-\hat{\varphi}$ is continuous, and the next proposition shows this.

Proposition X:1. Let $(h_i)_{i \in I}$ be a family of harmonic functions on U, uniformly bounded above. Then $\sup_{i \in I} h_i$ is continuous.

Proof. Without loss of generality, we can assume that all h_i, $i \in I$, are negative.

Put $H = \sup_{i \in I} h_i$. Since each h_i is continuous it is clear that H is lower semi-continuous so we have to prove that H also is upper semi-continuous. Let z_0 be given. If H is not upper semi-continuous at z_0 there is a sequence $(z_\nu)_{\nu=1}^\infty$ with limit z_0 such that $H(z_\nu) > H(z_0) + \varepsilon$ for some $\varepsilon > 0$. We can choose h_ν with

$$h_\nu(z_\nu) > H(z_0) + \varepsilon, \ \nu \in \mathbb{N}.$$

Take $r_0 > 0$ so that $B(z_0, r_0)$ is relatively compact in U and put

$$r_\nu = r_0 + |z_0 - z_\nu|.$$

We then have $\lim\limits_{\nu\to+\infty} r_\nu = r_0$ and $B(z_\nu, r_\nu)$, $\nu\in\mathbb{N}$, are relatively compact in U for ν large enough. Since

$$B(z_0, r_0) \subset B(z_\nu, r_\nu)$$

we have

$$\lim_{\nu\to+\infty} h_\nu(z_0) = \lim_{\nu\to+\infty} \frac{1}{m(B(z_0, r_0))} \int_{B(z_0, r_0)} h_\nu(z)dz =$$

$$= \lim_{\nu\to+\infty} \frac{m(B(z_\nu, r_\nu))}{m(B(z_0, r_0))} \frac{1}{m(B(z_\nu, r_\nu))} \int_{B(z_0, r_0)} h_\nu(z)dz \geq$$

$$\geq \lim_{\nu\to+\infty} \frac{m(B(z_\nu, r_\nu))}{m(b(z_0, r_0))} \cdot \frac{1}{m(B(z_\nu, r_\nu))} \int_{B(z_\nu, r_\nu)} h_\nu(z)dz =$$

$$= \lim_{\nu\to+\infty} \frac{m(B(z_\nu, r_\nu))}{m(B(z_0, r_0))} h_\nu(z_\nu) \geq H(z_0) + \varepsilon.$$

Hence, there is a ν so that

$$h_\nu(z_0) > H(z_0) + \varepsilon/2$$

which is a contradiction and the theorem is proved.

Example. Let $E = \{(z,0)\in\mathbb{C}^2; \ |z|=1\} \subset S = \partial B$ where B is the unit ball in $\mathbb{C}^2$. Then $R_{(z,\omega)}(E) = 0$, $\forall \ 0 \neq \omega$ (since E is polar). On the other hand, by Theorem X:1 and Lemma X:1, $Q_{(z,\omega)}(E) > 0$, $\forall (z,\omega)\in B$.

Lemma X:2. Let $(\varphi_i)_{i\in I}$ be a sequence of negative subharmonic functions on U. If there is a point $z_0\in U$ such that

$\inf_{i \in I} \varphi_i(z_0) > -\infty$ then there is a subharmonic function φ and a subsequence $(\varphi_{i_j})_{j=1}^{\infty}$ so that $\varphi_{i_j} \rightrightarrows \varphi$. Furthermore, if the functions φ_i, $i \in I$, are harmonic then h is harmonic and the convergence is uniform.

Proof. Since U is assumed to be connected, the condition $\inf_{i \in I} h_i(z_0) > -\infty$ implies that

$$\inf_{i \in I} \int_K h_i(z)\,dz > -\infty$$

for every compact subset K of U. Hence $(h_i)_{i \in I}$ contains a weakly convergent subsequence $(h_{i_j})_{j=1}^{\infty}$. If $\psi \in C_o^{\infty}(U)$, then $\lim_{j \to +\infty} \int h_{i_j} \Delta\psi \geq 0$ so the weak limit is a subharmonic function. This completes the proof, because it is a well-known fact that a weakly convergent sequence of harmonic functions converges uniformly on compacts.

Lemma X:3. If $\varphi = \sup_{\nu \in \mathbb{N}} \varphi_\nu$ where φ_ν are non-negative and upper semicontinuous on ∂U and if $\hat{\varphi}(0) < +\infty$ then $-\hat{\varphi}$ is continuous and plurisubharmonic on U.

Proof. By Lemma X:1, $-\widehat{\sup_{1 \leq \nu \leq m} \varphi_\nu}$ is plurisubharmonic and continuous for every m. Moreover

$$-\widehat{\sup_{1 \leq \nu \leq m} \varphi_\nu}(z) = -\sup_{\mu \in M_z} \int \sup_{1 \leq \nu \leq m} \varphi_\nu \, d\mu \searrow -\hat{\varphi}, \quad m \to +\infty$$

so since $-\hat{\varphi}(0) > -\infty$, $-\hat{\varphi}$ is plurisubharmonic on U.

It remains to prove continuity. We put $A=\{h\in\overline{P};\ h\leq-\hat{\varphi}\}$ and claim that A is non-empty and

$$\sup_{h\in A} h = -\hat{\varphi}$$

which will prove the continuity, by Proposition X:1.

By definition, $\sup_{h\in A} h \leq -\hat{\varphi}$ and Lemma X:1 provides us with a sequence $h_m\in P$ so that $h_m \leq -\widehat{\sup_{1\leq\nu\leq m} \varphi_\nu}$ and $-\frac{1}{m}-\hat{\varphi}(0) < h_m(0)$. Lemma X:2 shows that we can select a subsequence $(h_{m_j})_{j=1}^\infty$, converging uniformly on compact subsets. It is clear that this limit h_0 belongs to A and that $h_0(0)=-\hat{\varphi}(0)$. Thus we have shown that A is non-empty and that $-\hat{\varphi}>-\infty$ everywhere. So we can repeat the above argument for any point in U.

Theorem X:2. Let φ be a non-negative and universally measurable function on ∂U. If $\hat{\varphi}(0)<+\infty$ then there is an increasing sequence of upper semi-continuous functions φ_ν, $\nu\in\mathbb{N}$, such that $\varphi_\nu\leq\varphi$ and $\lim_{\nu\to+\infty} \hat{\varphi}_\nu=\hat{\varphi}$. Furthermore, $-\hat{\varphi}$ is plurisubharmonic and continuous on U.

Proof. Since φ is universally measurable we have

$$\hat{\varphi}(z)=\sup_{\mu\in M_z} \int\varphi(\xi)d\mu(\xi) = \sup_{g\leq\varphi}\ \sup_{\mu\in M_z} \int g(\xi)d\mu(\xi)$$

where g varies among the upper semi-continuous functions. Hence,

$$-\hat{\varphi}(z) =-\sup_{g\leq\varphi}\hat{g}(z) = \inf_{g\leq\varphi} -\hat{g}(z)$$

so $-\hat{\varphi}$ is upper semi-continuous by Lemma X:1.

By Choquets Lemma there is a denumerable subset $(g_j)_{j=1}^{\infty}$ of $\{g \leq \varphi\}$ so that if τ is lower semi-continuous and if $\tau \leq \inf_j -\hat{g}_j$ then $\tau \leq -\hat{\varphi}$. Now

$$-\sup_j g_j(z) = -\sup_{\mu \in M_z} \int \sup_j g_j d\mu \leq -\sup_{\mu \in M_z} \sup_j \int g_j d\mu \leq$$

$$\leq \inf_j -\sup_{\mu \in M_z} \int g_j d\mu = \inf_j -\hat{g}_j.$$

By Lemma X:3 the left-hand side is a continuous function so it follows that

$$-\widehat{\sup_j g_j}(z) \leq -\hat{\varphi}(z).$$

But on the other hand, $\widehat{\sup_j g_j} \leq \varphi$ so $\sup_j \hat{g}_j \leq \hat{\varphi}$ and therefore

$$-\hat{\varphi} = -\widehat{\sup_j g_j}.$$

If we take $\varphi_m = \sup_{1 \leq \nu \leq m} g_j$ we get a sequence of functions with the required properties.

Corollary X:1. If φ is a non-negative and universally measurable function on ∂U then $\hat{\varphi}(z) = \inf\{h(z); \ h \in \overline{P}, \ \hat{\varphi} \leq h\}$.

Proof. Use Theorem X:2 to find an increasing sequence $(\varphi_\nu)_{\nu=1}^{\infty}$ of upper semi-continuous functions such that $\varphi_\nu \leq \varphi$ and $\lim_{\nu \to +\infty} \hat{\varphi}_\nu = \hat{\varphi}$. Apply now the argument in the end of the proof of Lemma X:3.

Theorem X:3. Let E be a F_σ-set in ∂U of vanishing Q-capacity. Then there is a sequence $(f_\nu)_{\nu=1}^\infty$ of functions in $A(U)$ with $|f_\nu| \leq 2$ such that

1) $\lim\limits_{\nu \to +\infty} f_\nu(z) = 0, \quad z \in U,$

2) $\lim\limits_{\nu \to +\infty} f_\nu(\xi) = 0$ outside a set of vanishing Q-capacity,

3) $\lim\limits_{\nu \to +\infty} f_\nu(\xi) = 1$ on E.

Proof. We know that $E = \bigcup\limits_{\nu=1}^\infty K_\nu$ where $(K_\nu)_{\nu=1}^\infty$ is an increasing union of compact sets. By Lemma X:1, we can choose $g_\nu \in A(U)$, $\mathrm{Im}\, g_\nu(0) = 0$, $h_\nu = \mathrm{Re}\, g_\nu$ such that $h_\nu \geq \chi_{K_\nu}$; $h_\nu(0) < 1/\nu^3$. Put $P_\nu = e^{-\nu g_\nu}$. Then $|P_\nu| \leq 1$ since $h_\nu \geq 0$ and if $z \in E$ we get

$$\lim\limits_{\nu \to +\infty} P_\nu(z) = \lim\limits_{\nu \to +\infty} e^{-\nu h_\nu(z)} = 0.$$

Furthermore,

$$0 \leq \int (1 - \mathrm{Re}\, P_\nu)\, d\mu = 1 - \mathrm{Re}\, P_\nu(0) = 1 - e^{-\nu h_\nu(0)} < 1/\nu^2$$

for every $\mu \in M_0$ and every $\nu \in N$. It follows that $\lim\limits_{\nu \to +\infty} \mathrm{Re}\, P_\nu(z) = 1$ a.e. (μ) so $\lim\limits_{\nu \to +\infty} P_\nu(z) = 1$ a.e. (μ) since $|P_\nu| \leq 1$.

Since μ was any measure in M_0 it follows that $\lim\limits_{\nu \to +\infty} P_\nu = 1$ outside a set of Q-capacity zero. If we put $f_\nu = 1 - P_\nu$ we get a sequence of functions with property 2) and 3) and since $1 - \mathrm{Re}\, P_\nu \geq 0$; $1 - \lim\limits_{\nu \to +\infty} \mathrm{Re}\, P_\nu(0) = 0$, Lemma X:2 proves that

1-lim Re $P_\nu \equiv 0$ on U. Again, since $|P_\nu| \leq 1$,

$\lim\limits_{\nu \to +\infty} f_\nu(z) = \lim\limits_{\nu \to +\infty} 1 - P_\nu(z) \equiv 0$ on U, which completes the proof

of Theorem X:3.

Theorem X:4. Let φ be a non-negative subharmonic function

on the unit ball. Then φ has a pluriharmonic majorant if and

only if

$$\sup_{0<r<1} \sup_{\mu \in M_0} \int \varphi(r\xi) d\mu(\xi) < +\infty.$$

Proof. If $\varphi \leq h \in PH$ then $\varphi(r\xi) \leq h(r\xi)$ so

$$\sup_{0<r<1} \varphi(rg)(0) \leq h(0).$$

On the other hand, if φ is subharmonic then $\varphi(r\xi)\big|_{\partial B}$ is

upper semicontinuous. Hence, by Lemma X:1,

$$\widehat{\varphi(r\xi)}(z) = \inf\{h(z);\ h \in \text{Re } A(B),\ \varphi(r\xi) \leq h(\xi)\}$$

so choose $h_r \in \text{Re } A(B)$, $h_r(\xi) \geq \varphi(r\xi)$, with $\sup\limits_{0<r<1} h_r(0) < +\infty$.

Apply Lemma X:2 and choose a subsequence $(h_{r_j})_{j=1}^\infty$ con-

verging uniformly on compact subsets of B to h. Since φ

is subharmonic,

$$\varphi(r_j z) - h_{r_j}(z) \leq 0$$

on $\overline{B}$ so

$$\varphi(z_0) - h(z_0) = \lim_{j \to +\infty} \left[\varphi(r_j \frac{z_0}{r_j}) - h_{r_j}(\frac{z_0}{r_j}) \right] \leq 0,\ \forall z_0 \in B.$$

For the rest of this section we restrict ourselves to the unit ball although some of the results are true on more general domains.

Recall that $LH^1(B)$ is the Banach space of analytic functions with modulus having a plurisubharmonic majorant.

The norm on LH^1 is

$$\|f\|_{LH^1} = \inf\{h(0); \ h \in PH(B); \ |f| \leq h\}.$$

Observe that Theorem X:4 shows that $f \in H(B)$ is in LH^1 if and only if $\sup_{0<r<1} \widehat{|f(r\xi)|}(0) < +\infty$.

Theorem X:5. If $f \in LH^1$ then

$$\|f\|_{LH^1} = \lim_{r \to 1} \widehat{|f(r\xi)|}(0).$$

Proof. If $h \in PH$ with $h \geq |f|$ then $h(r\xi) \geq |f(r\xi)|$ so $h(0) \geq \widehat{|f(r\xi)|}(0)$. Hence

$$h(0) \geq \lim_{r \to 1} \widehat{|f(r\xi)|}(0) \geq \lim_{r \to 1} \widehat{|f(r\xi)|}(0)$$

which gives that

$$\|f\|_{LH^1} \geq \lim_{r \to 1} \widehat{|f(r\xi)|}(0).$$

On the other hand,

$$\lim_{r \to 1} \widehat{|f(r\xi)|}(z) \geq \int Q(z,\xi) \lim_{r \to 1} |f(r\xi)| d\mu(\xi) \geq |f(z)|$$

so Corollary X:1 proves that $\varlimsup\limits_{r\to 1} |f(r\xi)|(0) \geq \|f\|_{LH^1}$ which completes the proof.

In one variable, the functions in $A(B)$ are dense in H^1 (take dilatations).

Example. We are going to construct a bounded analytic function f on $B\subset\mathbb{C}^2$ such that

1) $\lim f(r\xi)$ exists $\forall\xi\in S$.

2) $f(r\xi)$ do not converge to f in LH^1.

Let $\xi_K = \left(\dfrac{1}{K}, \sqrt{1 - \dfrac{1}{K^2}}\right)$, $K\in\mathbb{N}$ Then $\{e^{i\Theta}\xi_K, \Theta\in\mathbb{R}\}$ are closed and disjoint sets so we can choose open disjoint sets V_K in $\overline{B}$ where V_k contains $\{e^{i\Theta}\xi_K, \Theta\in\mathbb{R}\}$.

For each K choose n_K so that

$$|<z,\xi_K>|^{n_K} < \frac{\varepsilon}{2^K}, \quad z\in\overline{B}\setminus V_K$$

and put $f(z)=\sum\limits_{K=1}^{\infty} <z,\xi_K>^{n_K}$. Then $|f(z)|\leq 1+\varepsilon$ since $z\in V_K$ for at most one K.

Let $0<r_1<1$ be given, choose K so that $r_1^{n_K}<1/4$ and then $0<r_2<1$ with $r_2^{n_K}>\dfrac{1}{2}$ such that $\lambda\in\mathbb{C}, |\lambda|=r_2 \Rightarrow \lambda\xi_K\subset V_K$.

For $w\in\mathbb{C}; |w|=1$ we now have

$$|f(wr_2\xi_K)-f(wr_1\xi_K)|=\left|\sum\limits_{p=1}^{\infty} <wr_2\xi_K,\xi_p>^{n_p}-<wr_1\xi_K,\xi_p>^{n_p}\right| =$$

$$= \left|\sum\limits_{p\neq K} <wr_2\xi_K,\xi_p>^{n_p} - <wr_1\xi_K,\xi_p>^{n_p} + w^{n_K}(r_2^{n_K} - r_1^{n_K})\right| >$$

$$> r_2^{n_K}-r_1^{n_K}-2\sum\limits_{p=1}^{\infty} \frac{\varepsilon}{2^p}\geq\frac{1}{4} - 2\varepsilon = \frac{1}{8} \quad \text{if} \quad \varepsilon = \frac{1}{16}.$$

Therefore $\sup\limits_{\mu \in M_0} \int |f(r_1\xi) - f(r_2\xi)| \, d\mu \geq \frac{1}{8}$ since we can take μ to be the normalized Lebesgue measure on $\{e^{i\Theta}\xi_K, \Theta \in \mathbb{R}\}$. This proves that f has property 2).

To prove 1) take $\xi \in S$ fix, then $f(r\xi) = \Sigma \, r^{n_k} \langle \xi, \xi_K \rangle^{n_K}$ with uniform convergence, since $\xi \in V_K$ for at most one K.

However, we have the following characterization of ALH^1, the functions in the closure of $A(B)$ with respect to the Lumer norm.

Theorem X:6. Assume that $f \in H(B)$. Then $f \in ALH^1(B)$ if and only if

$$Re f = \Sigma_\nu \, \varphi_1^\nu - \Sigma \, \varphi_2^\nu$$

and

$$Im \, f = \Sigma_\nu \, \psi_1^\nu - \Sigma \, \psi_2^\nu$$

where

$$\varphi_\mu^\nu, \ \psi_\mu^\nu, \quad \nu \in \mathbb{N}, \quad \mu = 1,2,$$

are non-positive and pluriharmonic on B, continuous up to the boundary.

Proof. $\Rightarrow$) Choose $f_n \in A(B)$ so that $\|f_n - f\|_{LH^1} \to 0$, $n \to +\infty$, and choose t_j so that

$$\|f_m - f_n\|_{LH^1} < 1/j^2, \quad m,n \geq t_j,$$

and put

$$f_{t_K} = f_{t_0} + \sum_{s=1}^{K} f_{t_s} - f_{t_{s-1}}.$$

Then

$$\left\| f_{t_s} - f_{t_{s-1}} \right\|_{LH^1} < \frac{1}{(s-1)^2}$$

so we can choose pluriharmonic majorants h_s with

$$h_s(0) < -\frac{1}{(s-1)^2}.$$

If we define

$$\varphi_1^\nu = \mathrm{Re}\ (f_{t_\nu} - f_{t_{\nu-1}}) - h_\nu, \quad \varphi_2^\nu = -h_\nu,$$

and

$$\psi_1^\nu = \mathrm{Im}\ (f_{t_\nu} - f_{t_{\nu-1}}) - h_\nu, \quad \psi_2^\nu = -h_\nu,$$

we get functions with the required properties.

$\Leftarrow$) Assume now that $f \in H(B)$ has the representation

$$f(z) = \sum_\nu \varphi_1^\nu - \sum_\nu \varphi_2^\nu + i\left(\sum_\nu \psi_1^\nu - \sum_\nu \psi_2^\nu \right).$$

Given $\varepsilon > 0$ choose N_ε so that

$$-\left(\sum_{\nu=N_\varepsilon}^{\infty} \varphi_1^\nu + \varphi_2^\nu + \psi_1^\nu + \psi_2^\nu \right)(0) < \varepsilon.$$

Then

$$|f(r\xi)-f(s\xi)| \leq \left| \sum_{\nu=1}^{N_\varepsilon} (\varphi_1^\nu(r\xi)-\varphi_1^\nu(s\xi)) - \sum_{\nu=1}^{N_\varepsilon} (\varphi_2^\nu(r\xi) - \right.$$

$$- \varphi_2^\nu(s\xi)) + i(\sum_{\nu=1}^{N_\varepsilon} \psi_1^\nu(r\xi)-\psi_1^\nu(s\xi)) - i(\sum_{\nu=1}^{N_\varepsilon} \psi_2^\nu(r\xi) -$$

$$\left. - \psi_2^\nu(s\xi)) \right| - \sum_{\nu=N_\varepsilon}^{\infty} (\varphi_1^\nu(r\xi)+\varphi_1^\nu(s\xi)+\varphi_2^\nu(r\xi)+\varphi_1^\nu(s\xi) +$$

$$+ \psi_1^\nu(r\xi)+\psi_1^\nu(s\xi)+\psi_2^\nu(r\xi)+\psi_2^\nu(s\xi)).$$

The first term at the right-hand side converges uniformly to zero when $r,s \nearrow 1$. The second sum is a pluriharmonic function which is less than 2ε at zero. Hence $(f(r\xi))_{0<r<1}$ is a Cauchy sequence in LH^1 and we have proved that $f(r\xi)$ tends to f in LH^1 and the proof is complete.

Remark. By Theorem X:6 we can to every $f \in ALH(B)$ define its boundary values: The term in the series $\sum_{\nu=1}^{\infty} \varphi_1^\nu$ are non-positive and continuous up to the boundary. If $\mu \in M_0$ then

$$\sum \varphi_1^\nu(0) = \int (\sum \varphi_1^\nu) d\mu$$

so $\sum \varphi_1^\nu(\xi) > -\infty$ outside a set (on ∂B) of vanishing Q-capacity. So we put

$$f^*(z) = \sum_\nu \varphi_1^\nu(\xi) - \sum_\nu \varphi_2^\nu(\xi) + i(\sum_\nu \psi_1^\nu(\xi) - \psi_2^\nu(\xi)), \quad \xi \in \partial B,$$

we get a function defined outside a set of vanishing Q-capacity. Furthermore, $\lim\limits_{r\to 1}\sup\limits_{\mu\in M_0}\int |f^*(\xi)-f(r\xi)|d\mu=0$.

We now use this last property to prove a result related to inner functions. Observe that we do not assume $|f|$ to be bounded.

Theorem X:7. $(n \geq 2)$ Assume that $f\in ALH^1(B)$ and that $\int |f^*|d\mu=1$, $\forall\mu\in M_0$. Then $f\equiv$ constant.

Proof. If $\mu=\sigma$ we get that

$$\int Q(z,\xi)|f^*(\xi)|d\sigma(\xi)\Big|_0 = 1$$

where Q is the classical Poisson kernel. To prove that $f\equiv$ const. it is enough to prove that $|f(0)|=1$ because that forces $|f|$ to be harmonic and therefore constant.

Choose for $0<r<1$ a measure $\mu_r\in M_0$ supported on

$$\{\xi\in\partial B;\ f(r\xi)=f(0)\}.$$

We have

$$|f(0)| = \lim_{r\to 1}\int |f(r\xi)|d\mu_r(\xi) \geq \lim_{r\to 1}\int |f^*|d\mu_r -$$

$$- \lim_{r\to 1}\int |f^*(\xi)-f(r\xi)|d\mu_r(\xi) = 1$$

since the last term vanishes by remark following Theorem X:6.

We now return to M_0; we have seen that to every function $f \in ALH^1$ there is a "boundary value function" $f*$ so that

1) $f(0) = \int f* d\mu, \quad \forall \mu \in M_0$

2) $\lim_{\substack{r \nearrow 1 \\ \mu \in M_0}} \sup \int |f*(\xi) - f(r\xi)| d\mu = 0.$

The above example shows that 2) need not hold if ALH^1 is replaced by $H^\infty(B)$. We do not know if 1) is true for $H^\infty(B)$. But $f*$ is well defined for every fixed $\mu \in M_0$!

A complex measure on S is called an A-<u>measure</u> if for every sequence, $f_j \in A(B)$, $|f_j(z)| \leq M$, $\forall j \in \mathbb{N}$, $\forall z \in B$ with $\lim_j f_j(z) = 0$ $\forall z \in B$ it follows that $\lim \int f_j d\mu = 0$.

Theorem X:8. If μ is an A-measure and if $f_j \in A(B)$, $|f_j(z)| \leq M$, $\forall z \in B$, $j \in \mathbb{N}$ such that $\lim_{j \to +\infty} f_j(z) \exists \forall z \in B$ then $f_j d\mu$ is weakly convergent (i.e. $\lim \int \varphi f_j d\mu \exists \forall \varphi \in C^0(S)$).

In particular, it follows from Theorem X:8 that if $\mu \in M_0$, $f \in H^\infty(B)$ then $f(r\xi) d\mu(\xi)$ is weakly convergent so there is a function (determined μ-a.e.) $f* \in L^\infty(S)$ such that $f(r\xi) d\mu(\xi) \rightharpoonup f* d\mu$.

Theorem X:9. Let μ be a probability-measure on S such that there is a constant c with

$$\sup_{0 < r < 1} \int |f(r\xi)|^2 d\mu(\xi) \leq c \int |f(\xi)|^2 d\mu(\xi), \quad \forall f \in A(B).$$

If $f \in H^\infty(B)$ then $(f(r\xi))_{0 < r < 1}$ converges in $L^2(\mu, S)$, $r \nearrow 1$.

<u>Notes and references</u>

Theorem X:3 is due to **F. Forelli**, Analytic measures. Pacific J. Math. 13 (163).

Theorem X:4 is a generalization of a theorem of **G. Lumer**, Espaces de Hardy en plusieurs variables complexes. C.R.A.S. Paris 273 (1971).

The example is a simplification of an example of **W. Rudin**, Function theory in the unit ball of $\mathbb{C}^n$. Springer Verlag 1980, pg. 150.

It has been proved by **W. Rudin** in "New constructions of functions holomorphic in the unit ball of $\mathbb{C}^n$", AMS regional conference series in mathematics No. 63 on page 64 that $ALH^1(B)$ contains no inner functions.

Theorem X:8 is due to **G.M. Henkin**, Banach spaces of analytic functions on the ball and on the bicylinder are not isomorphic. Functional Anal. Appl. 2 (1968).

Theorem X:9 is proved by **U. Cegrell** in On the strong convergence of dilatations of bounded analytic functions in the unit ball of $\mathbb{C}^n$. Manuscript. Toulouse. 1986.

For more references, see **U. Cegrell**, Small sets in $\mathbb{C}^n$, Ann. Pol. Math. (1985).

XI Szegö Kernels

We keep the notation from Section X.

1.   Approximation of the identity

Lemma XI:1. Let μ be a positive measure on $\partial\Omega$ and let

$(P_i(z,\xi))_{i=1}^{\infty}$, $(z,\xi)\in\partial\Omega\times\partial\Omega$ be a family of functions in

$L^1(\mu\otimes\mu,\ \partial\Omega\times\partial\Omega)$ such that

1)   a) $\int P_i(z,\xi)d\mu(\xi)=1$,   b) $\sup_{i,\xi}\int P_i(z,\xi)d\mu(z)=M<+\infty$

2) $P_i\geq0$, $i\in\mathbb{N}$

3) $P_i(z,\xi)d\mu(\xi)\rightsquigarrow\delta_z$, $i\rightarrow+\infty$, $\forall z\in\mathrm{supp}\ \mu$.

Then $\int P_i(z,\xi)\varphi(\xi)d\mu(\xi)$ tends to φ in $L^p(\mu,\partial\Omega)$, $i\rightarrow+\infty$ for

every $\varphi\in L^p(\mu,\partial\Omega)$.

Proof. Assume first that φ is continuous. Then

$$\lim_{i\rightarrow+\infty}\int P_i(z,\xi)\varphi(\xi)d\mu(\xi)=\varphi(z)$$

and since

$$\sup_{z,i}\left|\int P_i(z,\xi)\varphi(\xi)d\mu(\xi)\right|\leq\sup_{\xi}|\varphi(\xi)|$$

by 2) the lemma follows by dominated convergence. If $\varphi\in L^p$ and

if $\varepsilon>0$ is given, choose $\varphi_\varepsilon\in C(\partial\Omega)$ with $\int|\varphi-\varphi_\varepsilon|^p d\mu<\varepsilon/(M+1)$.

Then

$$\left(\int \left| \int P_i(z,\xi)\varphi(\xi)d\mu(\xi)-\varphi(z)\right|^p d\mu(z)\right) =$$

$$= \left(\int \left| \int P_i(z,\xi)(\varphi(\xi)-\varphi(z))d\mu(\xi)\right|^p d\mu(z)\right) \leq$$

$$\leq \left(\int\int P_i(z,\xi)\left|\varphi(\xi)-\varphi_\varepsilon(\xi)\right|^p d\mu(\xi)d\mu(z)\right) +$$

$$+ \left(\int\int P_i(z,\xi)\left|\varphi_\varepsilon(z)-\varphi(z)\right|^p d\mu(\xi)d\mu(z)\right) +$$

$$+ \left(\int \left| \int P_i(z,\xi)(\varphi_\varepsilon(\xi)-\varphi_\varepsilon(z))d\mu(\xi)\right|^p d\mu(z)\right).$$

Fubini´s theorem and property 1) show that the first two integrals are smaller than ε and the last integral tends to zero when $i\to+\infty$ by the first part of the proof. Since $\varepsilon>0$ was arbitrary, the lemma follows.

Lemma XI:2. Let $(P_i(z,\xi))_{i=1}^\infty$ be a family of functions as in Lemma XI:1. Assume that $(\psi_i)_{i=1}^\infty$ is a sequence of $L^1(\mu,\partial\Omega)$-functions such that $\psi_i d\mu \rightharpoonup \psi d\mu$ where $\psi\in L^1(\mu,\partial\Omega)$. If

$$\psi_i(z) \leq \int P_i(z,\xi)\psi(\xi)d\mu(\xi), \quad i\in\mathbb{N}, \quad z\in\partial\Omega,$$

then ψ_i tends to ψ in $L^1(\mu,\partial\Omega)$.

Proof. $\int \left|\psi(z)-\psi_i(z)\right|d\mu(z) \leq \int \left|\psi(z)-\int P_i(z,\xi)\psi(\xi)d\mu(\xi)\right|d\mu(z) +$

$+ \int\left(\int P_i(z,\xi)\psi(\xi)d\mu(\xi)-\psi_i(z)\right)d\mu(z).$

By Lemma XI:1, the first integral tends to zero when $i\to+\infty$. By weak convergence, we have the same conclusion for the second integral.

2.   Szegö and Poisson kernels

Let μ be a positive measure on $\partial\Omega$ such that to every compact subset K of Ω there is a constant C_K such that

$$(*) \qquad \sup_{z\in K} |f(z)| \leq C_K \left(\int_{\partial\Omega} |f|^2 d\mu\right)^{1/2}, \quad f\in A(\Omega).$$

Choose an ON-system $(e_\nu)_{\nu=1}^\infty$, $e_\nu \in A(\Omega)$ of functions dense in $A(\Omega)$ (and so in $H^2(\mu;\partial\Omega)$). Complete the system to an ON-basis $(e_\nu)_{\nu=1}^\infty \cup (d_\nu)_{\nu=1}^\infty$ in $L^2(\mu,\partial\Omega)$. Every $g\in L^2(\mu,\partial\Omega)$ then has a unique representation

$$g(z) = \sum_{\nu=0}^\infty \int g(\xi)\overline{e}_\nu(\xi)d\mu(\xi)e_\nu(z) +$$

$$+ \sum_{\nu=0}^\infty \int g(\xi)\overline{d}_\nu(\xi)d\mu(\xi)d_\nu(z), \quad z\in\partial\Omega.$$

For $m\in\mathbb{N}$,

$$\sum_{\nu=0}^m \int g\overline{e}_\nu d\mu e_\nu(z) \in A(\Omega)$$

so by $(*)$

$$\sup_{z\in K} \left|\sum_{m_1}^{m_2} \int g\overline{e}_\nu d\mu e_\nu(z)\right| \leq C_K \left\|\sum_{m_1}^{m_2} \int g\overline{e}_\nu d\mu e_\nu\right\| \to 0, \quad m_1,m_2 \to +\infty.$$

Thus $\sum_{\nu=0}^\infty \int g\overline{e}_\nu e_\nu$ converges uniformly on every compact subset of Ω and therefore represents an analytic function there.

If $f\in A(\Omega)$, we can apply the same argument on

$$f - \sum_{\nu=0}^{m} \int f\bar{e}_\nu d\mu e_\nu \quad \text{to conclude that}$$

$$f(z) = \sum_{1}^{\infty} \int f\bar{e}_\nu d\mu e_\nu(z), \quad z\in\Omega,$$

with uniform convergence on compact subsets of Ω.

If we consider the mappings

$$L^2(\mu) \ni g \overset{P}{\mapsto} \sum_{\nu=0}^{\infty} \int g\bar{e}_\nu d\mu e_\nu \in H^2(\mu)$$

and

$$H^2(\mu) \ni \sum_{1}^{\infty} \int g\bar{e}_\nu d\mu e_\nu \overset{V_z}{\to} \sum \int g\bar{e}_\nu d\mu e_\nu(z), \quad z\in\Omega,$$

it is clear that they are continuous so their composition $\tau_z = V_z \circ P$ is given by an element in $L^2(\mu)$:

$$V_z \circ P(g) = \int g\bar{\tau}_z d\mu, \quad g\in L^2(\mu).$$

It is easy to see that

$$\tau_z(\xi) = \sum_{\nu=1}^{\infty} \bar{e}_\nu(z)e_\nu(\xi), \quad \xi\in\partial\Omega$$

and we have seen that τ_z extends to an analytic function on Ω,

$$\tau_z(\omega) = \sum_{\nu=1}^{\infty} \bar{e}_\nu(z)e_\nu(\omega), \quad \omega\in\Omega.$$

Now, a crucial property of $(e_\nu)_{\nu=1}^{\infty}$ on Ω is: Are the e_ν:s linearly independent as analytic functions on Ω?

In other words: If $(\alpha_\nu)_{\nu=1}^\infty \in \ell^2$ and if $\Sigma \, \alpha_\nu e_\nu(z) \equiv 0$ on Ω must $\alpha_\nu = 0$, $\forall \nu \in \mathbb{N}$?

Theorem XI:1. Let μ be a positive measure satisfying property $(*)$. The following statements are equivalent.

1) If $f_\nu \in A(\Omega)$, $\int |f_\nu|^2 d\mu \leq 1$, $\nu \in \mathbb{N}$, and if $\lim_{\nu \to +\infty} f_\nu(z) = 0$, $\forall \nu \in \Omega$, then $f_\nu d\mu \not\to 0$.

2) There exists an ON-basis $(e_\nu)_{\nu=1}^\infty$ for $H^2(\mu)$ of functions in $A(\Omega)$, linearly independent on Ω.

Proof. 1) $\Rightarrow$ 2). If 2) is false, put $f_i(z) = \sum_{\nu=0}^{i} \alpha_\nu e_\nu(z)$.

2) $\Rightarrow$ 1). Assume that $f_i \in A(\Omega)$, $\int |f_j|^2 d\mu \leq 1$ and that $\lim_{\nu \to +\infty} f_\nu(z) = 0$, $z \in \Omega$.

Select a weakly convergent subsequence (which we again denote by (f_i) with limit $f \in H^2(\mu, \partial\Omega)$. Now

$$f_i(z) = \int f_i(\xi) \overline{\tau}_z(\xi) d\mu(\xi) \quad \text{so}$$

$$0 = \Sigma \, \left(\int f \overline{e}_\nu d\mu \right) e_\nu(z), \quad \forall z \in \Omega.$$

By assumption, $\int f \overline{e}_\nu d\mu = 0$, $\forall z \in \mathbb{N}$, which shows that the weak limit of $(f_i)_{i=1}^\infty$ is zero and the theorem is proved.

Corollary XI:1. Every positive measure on ∂B which satisfies $(*)$ and possesses a basis $(e_\nu)_{\nu=1}^\infty$ for H^2 so that

$$e_\nu(rz) = r^{g_\nu} e_\nu(z), \quad \nu, g_\nu \in \mathbb{N},$$

has the properties in the theorem.

Definition. The **Szegö kernel** (relatively μ and Ω) is

$$S(z,\xi) = \sum_{\nu=0}^{\infty} \overline{e}_\nu(z) e_\nu(\xi), \quad z \in \Omega, \; \xi \in \partial\Omega,$$

and the **Poisson kernel** is

$$P(z,\xi) = \frac{|S(z,\xi)|^2}{S(z,z)}, \quad z \in \Omega, \; \xi \in \partial\Omega.$$

It is clear that $P(z,\xi) \geq 0$. If $f \in A(\Omega)$, then

$$\partial\Omega \ni \xi \mapsto \frac{S(z,\xi)f(\xi)}{S(z,z)} \in H^2(\mu, \partial\Omega)$$

so

$$f(z) = \int \frac{S(z,\xi)f(\xi)}{S(z,z)} \; \overline{S(z,\xi)} d\mu(\xi) = \int P(z,\xi) f(\xi) d\mu(\xi).$$

Note that since P is real,

$$\mathrm{Re}\; f(z) = \int P(z,\xi) \mathrm{Re}\; f(\xi) d\mu(\xi), \quad f \in A(\Omega).$$

Theorem XI:2. Assume that the Choquet boundary of Ω relatively $A(\Omega)$ equals $\partial\Omega$ and let μ be a positive measure on $\partial\Omega$ having property (*). Furthermore, assume that there is a family $(F_i)_{i=1}^{\infty}$ of analytic mappings into Ω, each of them defined near $\overline{\Omega}$ such that $\lim\limits_{i \to +\infty} F_i(z) = z$, $\forall z \in \overline{\Omega}$.

If

$$\sup_{\substack{i \in \mathbb{N} \\ \eta \in \partial\Omega}} \int \frac{|S(F_i(z),\eta)|^2}{\int |S(F_i(z),\xi)|^2 d\mu(\xi)} \, d\mu(z) < +\infty$$

where S is the Szegö kernel relatively $\partial\Omega$ then μ has the following property. For every sequence $f_s \in A(\Omega)$, $|f_s| \leq 1$ with $\lim_{s \to +\infty} f_s(z) = 0$, $\forall z \in \Omega$ it follows that $f_s d\mu \rightarrow 0$, $s \to +\infty$.

Moreover, if $f \in H^\infty$ then $(f(F_i(\xi)))_{i=1}^\infty$ converges in $L^1(\mu, \partial\Omega)$.

Proof. Put $P_i(z,\xi) = P(F_i(z),\xi)$ where P is the Poisson kernel relative $\partial\Omega$. We first prove that $(P_i)_{i=1}^\infty$ is an approximation of the identity in the sense of Lemma 1.

First, each P_i is integrable on $\partial\Omega \times \partial\Omega$. Since

$$f(F_i(z)) = \int P_i(z,\xi) f(\xi) d\mu(\xi), \quad \forall f \in A(\Omega), \quad \forall z \in \overline{\Omega},$$

it is clear that $P_i(z,\xi) d\mu(\xi) \rightarrow \delta_z$, $i \to +\infty$ for every $z \in \partial\Omega$. (This is so because every point in $\partial\Omega$ is in the Choquet boundary.) Thus 3) of Lemma XI:1 is valid. It is trivial that 2) and the first part of 1) hold. We also have

$$1 = \int P_i(z,\xi) d\mu(\xi) = \int \frac{|S(F_i(z),\xi)|^2}{S(F_i(z),F_i(z))} \, d\mu(\xi).$$

Hence

$$S(F_i(z), F_i(z)) = \int |S(F_i(z),\eta|^2 d\mu(\eta)$$

so

$$\sup_{\substack{i\in\mathbb{N}\\ \xi\in\partial\Omega}} \int P_i(z,\xi)d\mu(z) = \sup_{\substack{i\in\mathbb{N}\\ \xi\in\partial\Omega}} \int \frac{|S(F_i(z),\xi)|^2 d\mu(z)}{S(F_i(z),F_i(z))} =$$

$$= \sup_{\substack{i\in\mathbb{N}\\ \xi\in\partial\Omega}} \int \frac{|S(F_i(z),\xi)|^2}{\int |S(F_i(z),\eta)|^2 d\mu(\eta)} \, d\mu(z) < +\infty$$

by assumption so the last part of 1) holds true.

Now, let $f_s \in A(\Omega)$, $|f_s| \leq 1$ be a given sequence with $\lim_{s\to+\infty} f_s(z)=0$, $\forall z\in\Omega$. If $f_s d\mu$ does not converge weakly to zero, then we can select a subsequence (which we again denote by $f_s d\mu$) such that $f_s d\mu \xrightarrow{\ \ } fd\mu$ where $0\not\equiv f\in L^\infty$.

We now wish to prove that

$$\lim_{s\to+\infty} \int P_i(z,\xi)f_s(\xi)d\mu(\xi) = \int P_i(z,\xi)f(\xi)d\mu(\xi), \qquad \forall z\in\Omega,$$

because then

$$\int P_i(z,\xi)f(\xi)d\mu(\xi) = \lim_{s\to+\infty} f_s(F_i(z))=0, \qquad \forall z\in\Omega,$$

by assumption. On the other hand, we have shown that Lemma XI:1 applies so $f=0$ a.e. (μ) which is a contradiction and the first part of the theorem would be proved.

So fix $i\in\mathbb{N}$ and $z\in\Omega$ and consider

$$S_j(F_i(z),\xi) = \sum_{\nu=0}^{j} \overline{e}_\nu(F_i(z))e_\nu(\xi).$$

Given $\varepsilon > 0$ choose j so that

$$\int |S_j(F_i(z),\xi) - S(F_i(z),\xi)|^2 d\mu(\xi) < \varepsilon S(F_i(z),F_i(z))$$

and then s so that

$$\left| \int |S_j(F_i(z),\xi)|^2 (f(\xi) - f_s(\xi)) d\mu(\xi) \right| < \varepsilon S(F_i(z),F_i(z)).$$

Then

$$\left| \int P_i(z,\xi) f(\xi) d\mu(\xi) - \int P_i(z,\xi) f_s(\xi) d\mu(\xi) \right| \leq$$

$$\leq \left| \int \left(\frac{|S(F_i(z),\xi)|^2}{S(F_i(z),F_i(z))} - \frac{|S_j(F_i(z),\xi)|^2}{S(F_i(z),F_i(z))} \right) (f(\xi) - \right.$$

$$\left. - f_s(\xi)) d\mu(\xi) \right| + \left| \int \frac{|S_j(F_i(z),\xi)|^2}{S(F_i(z),F_i(z))} (f(\xi) - f_s(\xi)) d\mu(\xi) \right| \leq$$

$$\leq 2 \int \frac{|S_j(F_i(z),\xi) - S(F_i(z),\xi)|^2}{S(F_i(z),F_i(z))} d\mu(\xi) + \varepsilon < 3\varepsilon$$

and the proof of the first part of the theorem is complete.

It remains to prove the last statement. If $f \in H^\infty$ then $f_i = f(F_i(z))$ is a uniformly bounded sequence in $A(\Omega)$. We can find a function $f \in L^\infty(\mu, \partial\Omega)$ and a sequence $(f_{i_j})_{j=1}^\infty$ so that $f_{i_j} d\mu \rightrightarrows g d\mu$ and by the proof above

$$\lim_{i \to +\infty} \int P_j(z,\xi) f(F_i(\xi)) d\mu(\xi) = \int P_j(z,\xi) g(\xi) d\mu(\xi).$$

Hence $f(F_j(z)) = \int P_j(z,\xi)g(\xi)d\mu(\xi)$ so another application of Lemma XI:1 completes the proof of the theorem.

3. $\underline{H^1(\sigma,\partial B) \text{ and } H^2(\sigma,\partial B)}$

We now return to the unit ball in B and σ, the normalized Lebesgue measure on ∂B.

It is clear that σ satisfies property $(*)$. The set $\left(\dfrac{z^\alpha}{c_\alpha^{1/2}}\right)_{\alpha \in \mathbb{N}^n}$ is an ON-basis for $H^2(\sigma,\partial B)$ where

$$c_\alpha = \int_{\partial B} |\xi^\alpha|^2 d\sigma(\xi) = \frac{(n-1)!\,\alpha!}{(n-1+|\alpha|)!}.$$

The Cauchy kernel is then

$$\overline{S}(z,\xi) = C[z,\xi] = \sum_\alpha \frac{z^\alpha\,\overline{\xi}^\alpha}{c_\alpha} = \frac{1}{(1-\langle z,\xi\rangle)^n}, \quad z\in B,\ \xi\in\partial B,$$

and the corresponding Poisson kernel

$$P(z,\xi) = \frac{|S(z,\xi)|^2}{S(z,z)} = \frac{(1-|z|^2)^n}{|1-\langle z,\xi\rangle|^{2n}}.$$

Observe that

$$\sup_{\substack{\xi\in\partial B \\ 0<r<1}} \int P(rz,\xi)d\sigma(z) = \sup_{\substack{\xi\in\partial B \\ 0<r<1}} \int \frac{(1-r^2)^n}{|1-\langle z,r\xi\rangle|^{2n}}\,d\sigma(z) = 1$$

so $P_r z(z,\xi) = P(rz,\xi)$ satisfies the conditions in Lemma XI:1

and the Theorem XI:2. Corollary XI:1 shows that σ has the equivalent properties in Theorem XI:1. This proves the first part in the following proposition.

Proposition XI:1. a) If $f_s \in A(B)$, $\int |f_s|^2 d\sigma \leq 1$, and if $\lim\limits_{s \to +\infty} f_s(z)=0$, $\forall z \in B$ then $f_s d\sigma \twoheadrightarrow 0$, $s \to +\infty$.

b) Let ν be the Lebesgue measure on B. The restriction map

$$L^1(\nu) \supset A(B) \ni f \xrightarrow{r} f\big|_{\partial B} \in H^1(\sigma, \partial B)$$

has a closed extension $\tilde{r}$ with

$$\text{Dom } \tilde{r} = \{f \in H(B); \sup_{0<r<1} \int |f(r\xi)| d\sigma(\xi) < +\infty\}$$

and

$$\text{Range } \tilde{r} = H^1(\sigma, \partial B).$$

Proof. We first prove (the well-known fact) that if $f \in H(B)$ and $\sup\limits_{0<r<1} \int |f(r\xi)| d\sigma < +\infty$ then $(f(r\xi))_{0<r<1}$ converges in $L^1(\sigma, \partial B)$.

Choose $\beta \in]0,1[$ and consider the uniformly integrable family $(|f(r\xi)|^\beta)_{0<r<1}$, $\xi \in \partial B$. Choose $r_\nu \nearrow 1$, $\nu \to +\infty$ and $\tau \in L^1(\gamma)$ so that $|f(r_\nu \xi)|^\beta d\sigma \twoheadrightarrow \tau d\sigma$. Let Q be the classical Poisson kernel for the unit ball in $\mathbb{R}^n$. Then

$$|f(z)|^\beta \leq \int Q(z,\xi) \tau(\xi) d\sigma(\xi), \quad z \in B,$$

$$- 127 -$$

where $\int Q\tau d\sigma$ is the smallest harmonic majorant of $|f|^{\beta}$.
Therefore τ is independent of $(r_{\nu})_{\nu=1}^{\infty}$ and Lemma XI:2 proves
that

$$|f(r\xi)|^{\beta} \to \tau, \quad r \nearrow 1 \quad \text{in} \quad L^{1}(\sigma).$$

By the Riesz-Fischer theorem, we can select a sequence
$s_{\nu} \nearrow 1$, $\nu \to +\infty$ so that $\lim_{\nu \to +\infty} |f(s_{\nu}\xi)|^{\beta} = \tau(\xi)$ a.e. (σ).

Fatou's lemma gives

$$\int \tau^{1/\beta} d\sigma = \int \lim_{\nu \to +\infty} |f(s_{\nu}\xi)| d\sigma \leq \varliminf_{\nu \to +\infty} \int |f(s_{\nu}\xi)| d\sigma(\xi) < +\infty$$

so $\tau^{1/\beta} \in L^{1}(\sigma)$ and

$$|f(z)| = (|f(z)|^{\beta})^{1/\beta} \leq \int Q(z,\xi)\tau(\xi)^{1/\beta} d\sigma(\xi)$$

by Jensen's inequality. Another application of Lemma XI:2 gives
the desired conclusion.

Now, if $f \in H(B)$ with $\sup_{0<r<1} \int |f(r\xi)| d\sigma(\xi) < +\infty$ we define
$\tilde{r}(f)$ as the $L^{1}(\sigma, \partial B)$-limit of $f(r\xi)$. If $f_{s} \to f$ in $L^{1}(\nu)$
and if $f_{s} \to g$ in $H^{1}(\sigma, \partial B)$ then it is clear that

$$\sup_{0<r<1} \int |f(r\xi)| d\sigma(\xi) < +\infty$$

so $\tilde{r}(f)$ is well defined. By Lemma XI:1,

$$\int P(rz,\xi)g(\xi)d\sigma(\xi) = \lim_{s \to +\infty} \int P(rz,\xi)f_{s}(g)d\sigma(\xi) = f(rz)$$

so $\tilde{r}(f)=\lim f(r\xi)=g(\xi)$ in $L^1(d\sigma)$ by Lemma XI:1. Thus $\tilde{r}$ is a closed operator and the proof is complete.

4. A-measures

In this section, we study extension problems related to Theorem XI:2. We also study their connection with the behaviour of the Szegö kernel relatively a certain measure.

Consider the restriction operator r,

$$L^\infty(dw) \supset H^\infty(\Omega) \supset A(\Omega) \xrightarrow{r} A(\Omega)\big|_{\partial\Omega} = A(\partial\Omega)$$

and let μ be a given measure on $\partial\Omega$. (Here dw is the Lebesgue measure on Ω.) We think of $A(\Omega)$ and $A(\partial\Omega)$ as subspaces of $(L^1(dw))'$ and $(L^1(\mu,\partial\Omega))'$ respectively and make the following definitions.

Definition. The measure μ is said to be <u>closed</u> if r has a closed extension $\tilde{r}$ in the weak*-topology.

Definition. We define $H^\infty(\mu,\partial\Omega)$ to be the weak*-closure of $A(\partial\Omega)$.

Definition. We say that Ω has property (**) if every $f \in H^\infty(\Omega)$ is the weak*-limit (relatively dw) of a sequence of functions in $A(\Omega)$.

Lemma XI:3. Consider the following conditions.

i) μ is a closed measure;

ii)   If $f_s \in A(\Omega)$, $s \in \mathbb{N}$, converges in the weak*-topology on Ω (relatively dw) to f, then $(f_s)_{s=1}^{\infty}$ is weak*-convergent in $L^1(\mu, \partial\Omega)$ and there is a uniformly bounded sequence $(g_s)_{s=1}^{\infty}$ in $A(\Omega)$ such that $\lim_{s \to +\infty} g_s(z)$ exist everywhere on Ω and equal to f and $\lim_{s \to +\infty} g_s(z)$ exist a.e. (μ) on $\partial\Omega$.

(iii) For every sequence $f_s \in A(\Omega)$, $s \in \mathbb{N}$, $\sup_{\substack{s \in \mathbb{N} \\ z \in \Omega}} |f_s(z)| \leq 1$ with $\lim_{s \to +\infty} f_s(z) = 0$, $\forall z \in \Omega$, we have

$$\lim_{s \to +\infty} \int \varphi(z) f_s(z) d\mu(z) = 0, \quad \forall \varphi \in C(\partial\Omega);$$

Then i) $\Rightarrow$ ii) $\Rightarrow$ iii) and if Ω has property (**) then iii) $\Rightarrow$ i).

Proof. i) $\Rightarrow$ ii). Assume that μ is closed and that $f_s \in A(\Omega)$, $s \in \mathbb{N}$, is a weak*-convergent sequence in $L^\infty(dw, \Omega)$. Since $L^1(dw, \Omega)$ is complete, $(f_s)_{s=1}^{\infty}$ is uniformly bounded on $\overline{\Omega}$. But since $L^1(\mu, \partial\Omega)$ is separable and complete, $(f_s)_{s=1}^{\infty}$ contains a subsequence that converges to a function f in the weak*-topology in $L^\infty(\mu, \partial\Omega)$. But the assumption that μ is closed now gives that the sequence $(f_s)_{s=1}^{\infty}$ itself converges to f in the weak*-topology. By the Banach-Saks theorem, we can select a subsequence $(f_{s_j})_{j=1}^{\infty}$ so that $g_m = \frac{1}{m} \sum_{j=1}^{m} f_{s_j}$ tends to f in $L^2(\mu, \partial\Omega)$ and by the Riesz-Fischer theorem, we can select another subsequence $(g_{m_j})_{j=1}^{\infty}$ so that $\lim_{j \to +\infty} g_{m_j} = f$ a.e. (μ). To complete the proof we observe that $(g_{m_j})_{j=1}^{\infty}$ is a uniformly bounded sequence since $(f_s)_{s=1}^{\infty}$ is.

That ii) $\Rightarrow$ iii) is clear.

iii) $\Rightarrow$ i). Assume that μ has property $(**)$. Thus, if $f \in H^{\infty}(\Omega)$ there is a sequence $f_s \in A(\Omega)$ converging to f in the weak*-topology. Since $L^1(dw, \Omega)$ is complete,

$$\sup_{\substack{s \in \mathbb{N} \\ z \in \Omega}} |f_s(z)| < +\infty$$

so, by ii),

$$\lim_{s \to +\infty} \int \varphi(z) f_s d\mu(z)$$

exists for every $\varphi \in L^1(\mu, \partial\Omega)$ and the limit is independent of the particular choice of the sequence $(f_s)_{s=1}^{\infty}$. We can now define $\tilde{r}(f)$ as this limit and it is clear that $\tilde{r}$ so defined is a closed operator.

Definition (cf. Section X). Let ν be a complex measure on $\partial\Omega$. Then ν is called an A-measure if for every uniformly bounded sequence $(f_s)_{s=1}^{\infty}$ of functions in $A(\Omega)$ with $\lim_{s \to +\infty} f_s(z) = 0$, $\forall z \in \Omega$, it follows that $\lim_{s \to +\infty} \int f_s d\nu = 0$.

Examples of A-measures are measures in M_0 and closed measures.

Theorem XI:3. Let μ be a regular (complex) Borel measure and M a weak*-compact and convex set of regular Borel probability measures on a compact Hausdorff spaces. Then μ has a unique decomposition

$$\mu = \mu_1 + \mu_2$$

where μ_1 is absolutely continuous with respect to a measure in M and μ_2 is carried by an F_σ-set E such that

$$\sup\{\mu(E);\ \mu\in M\} = 0.$$

Theorem XI:4. Let μ be an A-measure. Then

$$\mu = g d\nu + \eta$$

where $\nu\in M_0$, $g\in L^1(\nu)$ and η is carried by an F_σ-set of vanishing Q-capacity. Furthermore, $\eta\perp A(\Omega)$ (i.e. $\int f d\eta = 0$, $\forall f\in A(\Omega)$).

Proof. The set M_0 is weak*-compact. Use Theorem XI:3 to decompose $\mu = f d\nu + \eta$ and use Theorem X:3 to prove that $\eta\perp A(\Omega)$.

Corollary XI:2. Every closed measure is absolutely continuous with respect to a measure in M_0.

Proof. We know from Theorem XI:4 that if μ is closed then

$$\mu = f d\nu + \eta,\quad \nu\in M_0,\quad f\in L^1(\nu),\quad \eta\perp A(\Omega)$$

where the decomposition is unique. If $\varphi\in C(\partial\Omega)$ then $\varphi d\mu$ is closed so $\varphi d\eta$ $A(\Omega)$ by uniqueness. Hence $\int \varphi d\eta = 0$ for every continuous function φ so $\eta\equiv 0$ which proves the corollary.

Corollary XI:3. If μ is a positive A-measure then μ is absolutely continuous with respect to some measures in M_0.

Proof. By Theorem XI:3, $\mu=fd\nu+\eta$, $\eta\perp A(\Omega)$ and since μ is positive so is η. Thus $0=\int 1\cdot d\eta$ so $\eta\equiv 0$.

The example

$$\Omega = \{|z_1|<1\}\times\{|z_2|<1\}$$

and $\mu=\delta_1\otimes\delta_0-\delta_1\otimes\sigma_1$ (where σ_1 is the normalized Lebesgue measure on $|z_2|=1$) shows at the same time that A-measures need not to be closed or absolutely continuous with respect to any measure in M_0. To see this, take $\psi\in C_0^\infty(|z_2|<1)$, $\psi(0)\neq 0$, and consider

$$\int z_1^s\psi d\mu, \quad s\in\mathbb{N}.$$

Theorem XI:5. If Ω is strictly pseudoconvex then every A-measure is a closed measure.

Thus the equivalence of A-measures and closed measures depends of the shape of $\partial\Omega$. The next theorem shows that a relevant property of $\partial\Omega$ is the existence of a $\mu\in M_0$ such that its associated L^2-projection on $H^2(\mu,\partial\Omega)$ maps smooth functions on smooth functions.

Theorem XI:6. Assume that Ω has property (**) (cf. pg 128) and that there is a measure $\mu\in M_0$ with property (*) (cf. pg 118) such that μ has no point mass. Furthermore, assume that

1) μ is closed;

2) $S[z,\xi]$ is continuous on $\overline{\Omega}\times\partial\Omega\setminus\{z=\xi\}$;

3) $((\varphi(z)-\varphi(\xi)S(z,\xi))_{z\in\Omega}$ is a μ-uniformly integrable family

      for every $\varphi\in C^{\infty}(\mathbb{C}^n)$.

Then every   A-measure is a closed measure.

Proof.   Let $\varphi\in C^{\infty}(\mathbb{C})$. Property 2) and 3) show that

$$\left|\varphi(z)-\varphi(\xi)\right|\left\|S(z,\xi)\right|\in L^1(\mu,\partial\Omega)$$

for every $z\in\overline{\Omega}$.   Furthermore, by 3)

$$\overline{\Omega}\ni z \;\rightarrow\; \int\left|\varphi(z)-\varphi(\xi)\right|\left\|S(z,\xi)\right\|f(\xi)\left|d\mu(\xi)\right.$$

and

$$\overline{\Omega}\ni z \;\rightarrow\; \int(\varphi(z)-\varphi(\xi)\overline{S}(z,\xi)f(\xi)d\mu(\xi)$$

is continuous for every $f\in L^{\infty}(\mu,\partial\Omega)$.   Assume that $(f_i)_{i=1}^{\infty}$ is a uniformly bounded sequence of functions in $A(\Omega)$ with $\lim\limits_{i\to+\infty} f_i(z)=0$, $\forall z\in\Omega$,   and let $\varphi\in C(\partial\Omega)$ be given. We want to prove that $\lim\limits_{i\to+\infty}\int f_i\varphi d\nu=0$ for every   A-measure ν and since $C^{\infty}(\mathbb{C}^n)$ is dense in $C(\partial\Omega)$ we can assume that $\varphi\in C^{\infty}(\mathbb{C}^n)$.

     Put

$$I_i(z)=\int f_i(\xi)(\varphi(z)-\varphi(\xi))\overline{S}(z,\xi)d\mu(\xi)$$

and

$$J_i(z) = \int f_i(\xi)\varphi(\xi)\overline{S}(z,\xi)d\mu(\xi), \quad z\in\Omega.$$

We have already seen that $I_i(z)$ extends to a continuous function on $\overline{\Omega}$ and since

$$f_i(z)\varphi(z) = I_i(z) + J_i(z)$$

so does $J_i(z)$, and moreover,

$$\sup_{\substack{i\in\mathbb{N}\\ z\in\Omega}} |I_i(z)| + |J_i(z)| < +\infty.$$

We have assumed that μ is a closed measure, so since, for $z\in\overline{\Omega}$, $(\varphi(z)-\varphi(\xi))\overline{S}(z,\xi)\in L^1(\mu,\partial\Omega)$ it follows that $\lim_{i\to+\infty} I_i(z)=0$ and dominated convergence gives that $\lim_{i\to+\infty} \int I_i d\nu=0$ for every measure on $\partial\Omega$. Furthermore, $J_i\in A(\Omega)$ and $\lim_{i\to+\infty} J_i(z)=0$, $\forall z\in\Omega$, again since μ is closed. Thus $\lim_{i\to+\infty} \int J_i d\nu=0$ for every A-measure ν which means that

$$\lim_{i\to+\infty} \int \varphi f_i d\nu = \lim_{i\to+\infty} \int I_i d\nu + \lim_{i\to+\infty} \int J_i d\nu = 0$$

which proves the theorem.

Corollary XI:4. Assume that there is a measure μ on $\partial\Omega$ with the properties in Theorem XI:6 and that every point has vanishing Q-capacity. Then $H^\infty(\mu,\partial\Omega)+C(\partial\Omega)$ is a closed sub-algebra of $L^\infty(\mu,\partial\Omega)$.

Proof. Assume first that $f\in H^\infty(\mu,\partial\Omega)$ and $\varphi\in C^\infty(\mathbb{C}^n)$. Then

$$\varphi(z)f(z)=\int(\varphi(z)-f(\xi))\overline{S}(z,\xi)f(\xi)d\mu(\xi)+\int\varphi(\xi)f(\xi)\overline{S}(z,\xi)d\mu(\xi), \quad z\in\Omega,$$

and by the proof of Theorem XI:6 it follows that the first term extends continuously to $\overline{\Omega}$ while the second is in $H^\infty(\mu,\partial\Omega)$. By approximation, it remains to prove that $H^\infty(\mu,\partial\Omega)+C(\mu\Omega)$ is closed and this follows from ii) in the next theorem.

Theorem XI:7. Assume that Ω has property $(**)$ and that every point in $\partial\Omega$ has vanishing Q-capacity and that every A-measure is a closed measure. Then

i) if $\mu\in M_0$ with property $(*)$. Then

$$H^\infty(\mu+|\nu|,\partial\Omega) \xrightarrow{\ j\ } H^\infty(\mu,\partial\Omega)$$

is an isometry for every A-measure ν;

ii) if $\mu\in M_0$ with property $(*)$ then $H^\infty(\mu,\partial\Omega)+C(\partial\Omega)$ is a closed subset of $L^\infty(\mu,\partial\Omega)$.

Proof. i) It is clear that j is a continuous map from a Banach algebra into a Banach algebra so it is enough to prove that j is bijective. First supp $\mu=\partial\Omega$ for otherwise, by $(*)$, the Choquet boundary of $\partial\Omega$ (with respect to $A(\Omega)$) would be strictly smaller than $\partial\Omega$ and there would then be a positive measure Θ on $\partial\Omega\setminus\{\xi_0\}$ so that $\Theta-\delta_{\xi_0}\perp A(\Omega)$ for a point $\xi_0\in\partial\Omega$ and since every A-measure is assumed to be a closed measure this would imply that δ_{ξ_0} is a closed measure. Corollary XI:2 shows that the Q-capacity of $\{\xi_0\}$ is strictly positive, a contradiction.

Hence, the sup norm and ess sup (μ) are equal for continuous functions on $\partial\Omega$. So if $(f_s)_{s=1}^{\infty}$ is a sequence of functions in $A(\Omega)$ converging weak* in $L^{\infty}(\mu,\partial\Omega)$ then the family is uniformly bounded on Ω and it follows now from Lemma XI:3 and property (*) that $(f_s)_{s=1}^{\infty}$ converges weak* in $L^{\infty}(dw,\Omega)$ and so in $L^{\infty}(\mu+|\nu|,\partial\Omega)$ since $\mu+|\nu|$ is closed by assumption. It follows that j is a bijection which completes the proof of i).

ii) By the above part of the proof we can apply Theorem 2 and Proposition 1 in Aytuna and Chollet [1], which completes the proof of the theorem.

We finish this section with two examples where Theorem XI:6 and Corollary XI:4 can be applied.

Example 1. We choose $\Omega=B$, the unit ball in $\mathbb{C}^n$ and $\mu=\sigma$, the normalized Lebesgue measure on ∂B. Then Ω is strictly pseudoconvex and Theorem XI:5 shows that every A-measure is a closed measure. However, we want to show that the conditions in Theorem XI:6 are satisfied. (Hence, Corollary XI:4 applies in this case.) We know that

$$\overline{S}(z,\xi) = C[z,\xi] = \frac{1}{|1-<z,\xi>|^n}.$$

Thus property 2) in Theorem XI:6 is satisfied and by Proposition XI:1 we know that σ is a closed measure, so 1) holds true.

It remains to understand that 3) holds true. Choose $\alpha \in]1, \frac{n}{n-1/2}[$ and consider

$$I(z)=\int_{\partial B} \left(\frac{|z-\xi|}{|1-\langle z,\xi\rangle|^n}\right)^{\alpha} d\sigma.$$

If $\sup_{z\in B} I(z)<+\infty$ then it is clear that 3) holds. But

$$\frac{|z-\xi|}{|1-\langle z,\xi\rangle|^n} = \frac{\langle z-\xi,z-\xi\rangle^{1/2}}{[(1-\mathrm{Re}\langle z,\xi\rangle)^2+(\mathrm{Im}\langle z,\xi\rangle)^2]^{n/2}} \leq$$

$$\leq 2^{1/2} \frac{(1-\mathrm{Re}\langle z,\xi\rangle)^{1/2}}{[(1-\mathrm{Re}\langle z,\xi\rangle)^2+(\mathrm{Im}\langle z,\xi\rangle)^2]^{n/2}} \leq \frac{2}{|1-\langle z,\xi\rangle|^{n-(1/2)}}$$

By Rudin [7, Proposition 1.4.10] $\sup_{z\in B} I(z)<\infty$.

Example 2. Let $\alpha_1,\ldots,\alpha_n$ be n numbers in the interval $]0,1[$ and put

$$\Omega = D_{\alpha} = \{z\in\mathbb{C}^n;\ \sum_{j=1}^{n} |z_j|^{2/\alpha_j}<1\}.$$

Then D_{α} is a pseudoconvex set with C^2-boundary, but as soon as at least one α_j is less than 1, D_{α} is not strictly pseudoconvex.

We shall now see that the conditions in Theorem XI:6 are satisfied for a certain measure μ that we define inductively. For $n=1$ we define $\mu=\sigma_1=$ the normalized Lebesgue measure on $|z|=1$.

Assume now that $\mu_{\alpha'}$ is defined for all $\alpha'=(\alpha_1,\ldots,\alpha_{n-1})$.

We then define μ_α for $\alpha=(\alpha',\alpha_n)$ as

$$\int_{\partial D_\alpha} f(z)d\mu_\alpha = \int_{\partial D_{\alpha'}\times\{|z_n|<1\}} f((1-|z_n|^{2/\alpha_n})^{1/2}z',z_n) \cdot$$

$$\cdot (1-|z_n|^{2/\alpha_n})^{|\alpha'|-1} d\mu_{\alpha'} dz_n,$$

where dz_n denotes the Lebesgue measure on $\{|z_n|<1\}$. It is clear that $\mu_\alpha \in M_0$ and that μ_α is closed is easily seen from the fact that σ_1 is closed.

For μ_α, Bonami and Lohoué have calculated the corresponding Szegö kernel S_{μ_α} and they prove in Proposition 2:1 that it extends to an infinitely differentiable function on $\overline{D}_\alpha \times \overline{D}_\alpha$ outside the diagonal of $\partial D_\alpha \times \partial D_\alpha$. (Bonami and Lohoué[2].)

Furthermore, they prove in Proposition 5.3 that $S_{\mu_\alpha}(z,\xi)|z-\xi|$ is uniformly integrable in $L^p(\mu_\alpha)$ for every

$$p < \frac{1}{1 - \frac{1}{2n}\inf\alpha_k} \quad \text{which shows that}$$

$$((\varphi(z)-\varphi(\xi))S_{\mu_\alpha}(z,\xi))_{z\in D_\alpha}$$

is a μ_α-uniformly integrable family for every Lipschitz function φ. It follows that the conditions of Theorem XI:6 are satisfied and we can also apply Corollary XI:4.

5.   The Cauchy transform of measures

We consider again the unit ball B in $\mathbb{C}^n$ and σ, the normalized Lebesgue measure on ∂B. We write C and P for the corresponding Cauchy and Poisson kernel, and we say that a function $f \in H(B)$ belongs to H^p, if $\tilde{r}(f) \in H^p(\sigma, \partial B)$. If

$$g(z) = \int_{\partial B} P(z, \xi) d\mu(\xi) \in H(B)$$

for a complex measure μ then $\mu = f d\mu$ where $f \in H(\sigma, \partial B)$ and supp $\mu = \partial B$ (or empty). This is so because

$$\int |g(rz)| d\sigma(z) = \int |\int P(rz, \xi) d\mu(\xi)| d\sigma(z) \leq \int d|\mu|(\xi) < +\infty$$

so $g \in H^1$ and $(g(r\xi))_{0<r<1}$ converges in $L^1(\sigma, \partial B)$ to g^* (Proposition XI:1). On the other hand,

$$g(rz) = \int P(rz, \xi) d\mu(\xi)$$

converges weakly to μ so $\mu = g^* d\sigma$.

Recall now that the Cauchy kernel relatively B and σ is

$$C[z, \xi] = \frac{1}{(1-\langle z, \xi \rangle)^n} = \sum_{\alpha \geq 0} \frac{\bar{z}^\alpha \xi^\alpha}{c_\alpha} \, ,$$

$z \in B$, $\xi \in \partial B$. If μ is any measure on ∂B for which the functions (ξ^α) are pairwise orthogonal then

$$\int C[z, \xi] d\mu(\xi) = \mu(1).$$

For instance, let μ be defined on the unit sphere in $\mathbb{C}^n$ by

$$\int_{\partial B} \varphi(\xi)d\mu(\xi) = \frac{1}{2\pi} \int_0^{2\pi} \varphi(e^{i\Theta},0)d\Theta$$

(i.e. $\mu=\sigma\otimes\delta_0$). Then μ is singular relatively σ but the Cauchy transform of $\mu \equiv 1$. But there are restrictions on μ in order to have

$$\int C[z,\xi]d\mu \in H^1 .$$

Theorem XI:8. Let μ be a real measure so that $g(z)=\int C[z,\xi]d\mu(\xi)\in H^1$. Then μ is a closed measure.

Proof. Let $f_i \in A(B)$, $i\in\mathbb{N}$, be a uniformly bounded sequence such that $\lim\limits_{i\to+\infty} f_i(z)=0$, $\forall z\in B$. Then

$$\lim_{i\to+\infty} \int f_i(\xi)d\mu(\xi) = \lim_{i\to+\infty} \lim_{r\to 1} \int (\int C[rz,\xi]f_i(z)d\sigma(z))d\mu(\xi)=$$

$$= \lim_{i\to+\infty} \int f_i(z)\overline{g^*}(z)d\sigma(z) = 0$$

by Proposition XI:1 where g^* is the $L^1(\sigma)$-limit of $g(rz)$, $r\nearrow 1$, which exists since $g\in H^1$.

We have proved that μ is an A-measure and we know that every A-measure is a closed measure.

Corollary XI:4. If $\int C[z,\xi]d\mu(\xi)\in H^1$ for a real measure $\mu(\neq 0)$ then the Q-capacity of the support of μ is positive.

Proof. By Theorem XI:8 and Corollary XI:2, $\mu = f d\nu$ where $\nu \in M_0$ and $f \in L^1(\nu, \partial\Omega)$. Thus $\text{supp } \mu = (\text{supp } \nu) \cap (\overline{f \neq 0})$.

Remark. There is no converse of Theorem XI:8. Choose $n = 1$ and a non-negative function $\tau \in L^1(\sigma, \partial B) \setminus L \log L$. Then $\tau d\sigma$ is a closed measure but $g(z) = \int C[z, \xi] \tau d\sigma \notin H^1$ because if $g \in H^1$, $\text{Re } g \leq 0$ then $\text{Re } g^* \in L \log L$. Now, since $2 \text{ Re } C - 1 = P$ we get that

$$2 \text{ Re } g^* - \int \tau(\xi) d\sigma(\xi) = \tau,$$

a contradiction, since the left-hand side is in $L \log L$.

For the rest of this section, let μ denote a positive measure so that the functions $(\xi^\alpha)_{\alpha \geq 0}$ are pairwise orthogonal and remember that C denotes the Cauchy kernel relative to B and σ.

By construction, if $\varphi \in L^2(\sigma, \partial B)$ then

$$\sum_{\alpha \geq 0} \int \varphi(\xi) \overline{\xi}^\alpha d\sigma(\xi) \, \frac{z^\alpha}{c^\alpha}$$

is the L^2-projection of φ into H^2. The following theorem shows what happens when we replace σ by μ.

Theorem XI:9. The map $L^2(\mu, \partial B) \ni \tau \to \int C[z, \xi] \tau(\xi) d\mu(\xi)$ is into H^2 if and only if $\sup_{\alpha \geq 0} \frac{d_\alpha}{c_\alpha} < +\infty$ where $d_\alpha = \int |\xi^\alpha|^2 d\mu(\xi)$ and $c_\alpha = \int |\xi^\alpha|^2 d\sigma(\xi)$. Furthermore, if $\sum d_{\alpha_j} = +\infty$ for a sequence

$(\alpha_j)^\infty_{j=1}$ with $\sum\limits_{j=1}^\infty (\dfrac{c_{\alpha_j}}{d_{\alpha_j}})^{1/2} < +\infty$ then there is a continuous

function τ on ∂B such that

$$\int C[z,\xi]\tau(\xi)d\mu(\xi) \notin H^2.$$

(Remember that we have assumed that $\int C[z,\xi]d\mu(\xi) \equiv \mu(1) > 0$.)

Proof. If $\varphi \in L^2(\mu,\partial B)$ then

$$\varphi(z) = \sum_{\alpha \geq 0} \int \frac{\varphi(\xi)\overline{\xi}^\alpha}{d_\alpha^{1/2}} d\mu(\xi) \frac{z^\alpha}{d_\alpha^{1/2}} + P(z)$$

where

$$\int P(z)\overline{\xi}^\alpha d\mu(z) = 0, \quad \forall \alpha \in \mathbb{N}^n.$$

Hence

$$\int C[z,\xi]\varphi(\xi)d\mu(\xi) = \int \sum \frac{v^\alpha \overline{\xi}^\alpha}{c_\alpha} \varphi(\xi)d\mu(\xi) =$$

$$= \sum \int \frac{\varphi(\xi)\overline{\xi}^\alpha}{c_\alpha^{1/2}} d\mu(\xi)\frac{z^\alpha}{c_\alpha^{1/2}} = \sum \frac{d_\alpha^{1/2}}{c_\alpha^{1/2}} \int \frac{\varphi(\xi)\overline{\xi}^\alpha}{d_\alpha^{1/2}} d\mu(\xi)\frac{z^\alpha}{c_\alpha^{1/2}}$$

so if $\sup\limits_\alpha \dfrac{d_\alpha}{c_\alpha} < +\infty$ then $\int C[z,\xi]\varphi(\xi)d\mu(\xi) \in H^2$. Assume now that

$\sup\limits_\alpha \dfrac{d_\alpha}{c_\alpha} = +\infty$ and choose a sequence $(\alpha_j)^\infty_{j=1}$ such that

$\sum\limits_{j=1}^\infty (\dfrac{c_{\alpha_j}}{d_{\alpha_{jk}}})^{1/2} < +\infty$ and consider

$$\varphi(\xi) = \sum_{j=1}^{\infty} \frac{c_{\alpha_j}^{1/4}}{d_{\alpha_j}^{3/4}} \, \xi^{\alpha_j} = \sum_{j=1}^{\infty} \frac{c_{\alpha_j}^{1/4}}{(d_{\alpha_j})^{1/4}} \, \frac{\xi^{\alpha_j}}{d_{\alpha_j}^{1/2}} \ .$$

Then

$$\int |\varphi|^2 d\mu = \sum_{j=1}^{\infty} \left(\frac{c_{\alpha_j}}{d_{\alpha_j}}\right)^{1/2} > +\infty$$

but

$$\int C[z,\xi]\varphi(\xi)d\mu(\xi) = \sum_{j=1}^{\infty} \left(\frac{d_{\alpha_j}}{c_{\alpha_j}}\right)^{1/4} \frac{z^{\alpha_j}}{c_{\alpha_j}^{1/2}}$$

which is not in H^2 since $\lim\limits_{j \to +\infty} \dfrac{d_{\alpha_j}}{c_{\alpha_j}} = +\infty$.

Finally, assume that

$$\sum_{j=1}^{\infty} \left(\frac{c_{\alpha_j}}{d_{\alpha_j}}\right)^{1/2} < +\infty \quad \text{but} \quad \sum_{j=1}^{\infty} d_{\alpha_j} = +\infty.$$

Then

$$\psi(z) = \sum_{j=1}^{\infty} \left(\frac{c_{\alpha_j}}{d_{\alpha_j}}\right)^{1/2} z^{\alpha_i}$$

is continuous on ∂B but

$$\int C[z,\xi]\psi(\xi)d\mu(\xi) = \sum_{j=1}^{\infty} d_{\alpha_j} \frac{c_{\alpha_j}^{1/2}}{d_{\alpha_j}^{1/2}} \frac{z^{\alpha_j}}{c_{\alpha_j}} = \sum_{j=1}^{\infty} \left(\frac{d_{\alpha_j}}{c_{\alpha_j}}\right)^{1/2} z^{\alpha_j}$$

so

$$\int \left| \int C[z,\xi]\psi(\xi)d\mu(\xi) \right|^2 d\sigma(\xi) = \sum_{j=1}^{\infty} d_{\alpha_j} = +\infty$$

which completes the proof of Theorem XI:9.

Example. If $\nu=\sigma_1\times\delta_0$ then $\nu\in M_0$ and $\int \xi^\alpha \bar{\xi}^\beta d\nu=0$, $\alpha\neq\beta$. But $d(m,0)=1$, $\forall m\in\mathbb{N}$, so Theorem XI:9 shows that the continuous functions do not operate on the class T of measures

$$T = \{\eta;\ \text{supp}\ \eta \subset \partial B,\ \int C[z,\xi]d\eta(\xi)\in H^2\}.$$

Proposition XI:1 says that if $(f)_{s=1}^{\infty}$ is a sequence in $A(B)$, bounded in $H^2(\sigma,\partial B)$ and such that $\lim_{s\to+\infty} f_s(z)=0$, $\forall z\in B$ then $f_s d\sigma \not\rightarrow 0$, $s\to+\infty$.

We do not know if this can be extended to arbitrary measures in M_0 but we have the following theorem.

Theorem XI:10. Suppose that $(f_j)_{j=1}^{\infty}$ is a sequence of functions in $A(B)$ such that

$$\sup_j \int |f_j|^{2n+\epsilon}d\sigma < +\infty$$

and that $\lim_{j\to+\infty} f_j(z)$ exists for every $z\in B$. If $\mu\in M_0$ then $f_j d\mu$ is convergent in the sense of distribution theory. If furthermore $\sup_j \int |f_j|d\mu < +\infty$ then $f_j d\mu$ is weakly convergent.

Proof. Suppose that $(f_j)_{j=1}^{\infty}$ satisfies the assumptions in the theorem and that $\lim\limits_{j\to+\infty} f_j(z)=0$, $\forall z\in B$. We then have to prove that

$$\lim_{j\to+\infty} \int f_j\varphi d\mu = 0, \quad \forall\varphi\in C^\infty(\partial B)$$

where μ is fixed in M_0.

Assume that $\varphi\in C^1(\mathbb{C}^n)$ and let $0<r<1$. Then

$$\int f_j(rz)\varphi(rz)d\mu(z) = \int\varphi(rz)\int f_j(\xi)\overline{S}(rz,\xi)d\sigma(\xi)d\mu(z) =$$

$$= \int[\int f_j(\xi)(\varphi(rz)-\varphi(r\xi))\overline{S}(rz,\xi)d\sigma(\xi)]d\mu(z) +$$

$$+ \int f_j(\xi)\varphi(r,\xi)d\sigma(\xi)$$

where we have used that $\mu\in M_0$. It follows from Proposition XI:1 that $\lim\int f_j(\xi)\varphi(r_j\xi)d\sigma(\xi)=0$ for every sequence $r_j\nearrow 1$, $j\to+\infty$.

It remains to prove to take care of the first term. Note first that it follows from the calculation in Example 1, that

$$\sup_{w\in B} \int|\varphi(w)-\varphi(\xi)|^{\alpha'}|S(w,\xi)|^{\alpha'}d\sigma(\xi) < +\infty$$

where $\alpha=2n+\varepsilon$ and $\dfrac{1}{\alpha} + \dfrac{1}{\alpha'} = 1$.

Therefore, for $z\in\overline{B}$ fixed,

$$\lim_{j\to+\infty} \int f_j(\xi)(\varphi(r_jz)-\varphi(r_j\xi)\overline{S}(r_jz,\xi)d\sigma(\xi)=0$$

for every sequence $r_j \nearrow 1$, $j \to +\infty$. Thus, by dominated convergence, the first term at the right hand side tends to zero for every sequence $r_j \nearrow 1$, $1 \to +\infty$.

To finish the proof of Theorem XI:10, we need only observe that every continuous function on ∂B can be uniformly approximated by functions in $C^\infty(\mathbb{C}^n)$.

Notes and references

General references for this section are Bungart [3] and Rudin [7].

Theorem XI:3 was proved by a combination of arguments of Glicksberg, König, Seever and Rainwater. See Rudin [7, pg. 194] for details.

Theorem XI:5 is due to Henkin, see Henkin and Čirka [4].

We have no exact description of the sets that satisfies condition (**), (pg 94).

By taking dilatations, it is clear that every starshaped domain satisfies (**). It is known that this is true also for smooth strictly pseudoconvex domains, cf.: **N. Kerzman**; Hölder and L^p estimates for solutions of $\bar{\partial}u=f$ in strongly pseudoconvex domains, Comm. Pure Appl. Math. 24 (1971), 301-379, and **Brian Cole** and **R. Michael Range**; A-measures on complex manifolds and some applications, J. Func. An. 11 (1972), 393-400.

That $H^\infty+C$ is a closed subalgebra of L^∞ was first proved by Sarason [8] on the unit circle, by Rudin [6] on the unit sphere in $\mathbb{C}^n$, by Aytuna and Chollet [1] on the boundary of strictly pseudoconvex domains.

Jewell and Krantz [5] proved the results for convex sets in $\mathbb{C}^2$ with real analytic boundary and they also remarked that they had the result for $\partial D_\alpha \subset \mathbb{C}^n$ where $\alpha=(\alpha_1,\ldots,\alpha_n)$, $\frac{1}{\alpha_k} \in \mathbb{N}$, $1 \leq k \leq n$. The methods in the above papers are different from ours.

[1] **Aytuna, A and Chollet, A.-M.,** Une Extension d'un resultat de W. Rudin. Bull. Soc. Math. france 104 (1976), 383-388.

[2] **Bonami, A. and Lohoué, N.,** Projecteurs de Bergman et Szegö pour une classe de domains faiblement pseudo-convexes et estimations L^p. Compos. Math. 46 (1982), 159-226.

[3] **Bungart, L.,** Boundary kernel functions for domains on complex manifolds. Pacific J. Math 14 (1964), 1151-1164.

[4] **Henkin, G.M. and Čirka, E.M.,** Boundary properties of holomorphic functions of several complex variables. J. Soviet Math. 5 (1976), 612-687.

[5] **Jewell, N.P. and Krantz, S.G.,** Toeplitz operators and related function algebras on certain pseudoconvex domains. Trans. Amer. Math. Soc. 252 (1979), 297-312.

[6] **Rudin, W.,** Spaces of type $H^\infty+C$. Ann. Inst. Fourier 25 (1975), 99-125.

[7] **Rudin, W.,** Function theory in the unit ball of $\mathbb{C}^n$. Springer-Verlag. New York, Heidelberg, Berlin 1980.

[8] **Sarason, D.,** Generalized interpolation in H^∞. Trans. Amer. Math. Soc. 127 (1967), 179-203.

XII Complex Homomorphisms

In the last section, we saw that one could assign "boundary values" to certain analytic functions by considering closed extensions of the restriction operator.

Here we use the algebraic structure of $H^\infty(\Omega)$ and consider elements in $H^\infty(\Omega)$ as continuous functions on a compact set "containing" Ω. We start with a very brief review of the elements of Gelfand representation.

Let A be a uniform and commutative Banach algebra (with identity). A complex homomorphism (or linear multiplicative functional) on A is an element $0 \neq m \in A'$ such that

$$m(fg)=m(f)m(g), \quad \forall f,g \in A.$$

We write M_A for the complex homomorphisms on A. Note that M_A is contained in the unit sphere in A'.

Therefore, by the Banach-Alaoglu theorem, M_A is compact if we give it the weak*-topology, which is generated by neighborhoods of the form

$$\{m \in M_A; \ \max_{1 \leq j \leq k} |m(f_j)-m_0(f_j)| < \epsilon, \ f_j \in A \atop 1 \leq j \leq k \}$$

for $m_0 \in M_A$, $\epsilon > 0$.

One can show that every maximal ideal in A is the kernel of an element in M_A, therefore M_A is sometimes called the maximal ideal space of A.

With every element $f \in A$, we associate $\hat{f} \in C(M_A)$ by $\hat{f}(m) = m(f)$, $m \in M_A$ and it follows from the definition of the weak*-topology that $\hat{f}$ is always continuous on the compact space M_A.

If we denote by $\hat{A}$ the set of $\hat{f}$, $f \in A$ we have an algebra homomorphism $f \to \hat{f}$, $\sup\limits_{m \in M_A} |\hat{f}(m)| = \sup\limits_{m \in M_A} |m(f)| \leq \|f\|$ and $\hat{f}$ is called the Gelfand transform of f.

A closed subset K of M_A is called a boundary for A if

$$\|f\| = \sup\limits_{m \in K} |m(f)|, \quad \forall f \in A.$$

One can prove that there is a smallest boundary; this boundary is called the Shilov boundary of A.

Let now Ω be an open and bounded subset of $\mathbb{C}^n$, $n \geq 1$. Let A be a uniform Banach algebra of continuous functions on Ω (with supnorm). Then every $z_0 \in \Omega$ gives rise to an element $\tilde{z}_0 \in M_A$

$$\tilde{z}_0(f) = f(z_0).$$

We denote by $\pi(m) = (m(z_1), \ldots, m(z_n))$, $m \in M_A$, and $\overline{\Omega}^* =$ the weak*-closure of

$$\{m \in M_A; \ \pi(m) \in \Omega\}.$$

Proposition XII:1. The Shilov boundary of $H^\infty(\Omega)$ is contained in $\overline{\Omega}*$.

Proof. If $\sup\limits_{m\in\overline{\Omega}*}|\hat{f}(m)|<|\hat{f}(m_0)|$ for a $f\in H^\infty(\Omega)$ and a $m_0\in M_A$ then $\hat{f}-\hat{f}(m_0)$ and $1/f-\hat{f}(m_0)$ are in $H^\infty(B)$. Therefore

$$1=m_0(f-\hat{f}(m_0))m_0(\frac{1}{f-\hat{f}(m_0)})=0$$

which is a contradiction.

We will need the following refinements of Lemma III:2 and Theorem X:3.

Theorem XII:1. Let G be a convex subset of some vectorspace, K a convex subset of some topological vectorspace and $F: G\times K\to\mathbb{R}$ a function such that $G\ni x\mapsto F(x,y)$ is convex on G for every $y\in K$. $K\ni y\mapsto F(x,y)$ is concave and continuous on K for every $x\in G$. Then $\sup\limits_{y\in K}\inf\limits_{x\in G}F(x,y)=\inf\limits_{x\in G}\sup\limits_{y\in K}F(x,y)$.

To formulate the next theorem, we need some more notation. Let X be a compact Hausdorff-space, $M(X)$ is the set of all regular Borel measures on X. A function algebra A on X is a closed subalgebra of $C(X)$ (with sup-norm). We assume that A contains all the constants and separates points on X. If h is a multiplicative linear functional on A, then it follows from the Hahn-Banach theorem and Riesz representation theorem that there exists a probability measure $\rho\in M(X)$ such that

$$h(f) = \int f d\rho, \quad \forall f \in A$$

and we then say that ρ represents h. We define M_h to be all the probability measures that represents h.

Theorem XII:2. With notation as above, suppose E is an F_σ-set in X such that $\sup\limits_{\rho \in M_h(X)} \rho(E)=0$. Then there is a sequence $\{f_m\}_{m=1}^\infty$ of functions in A such that

$$\|f_m\| \leq 1, \quad \forall m \in \mathbb{N}$$

$$\lim_{m \to +\infty} f_m(x)=0, \quad \forall x \in E$$

$$\lim_{m \to +\infty} f_m(x)=1 \quad \text{a.e. } (\rho), \quad \forall \rho \in M_h.$$

Let now B be the unit ball in $\mathbb{C}^n$, $n \geq 1$, $A=H^\infty(B)$. Then $H^\infty(B)$ is a closed subalgebra of $C(M_A)$ and by the Stone-Weierstrass theorem, $(\widehat{f\overline{g}})_{f,g \in H^\infty(B)}$ generates $C(M_A)$. Therefore, if $\mu \in M_0$ then

$$\widehat{f\overline{g}} \mapsto \int f^* \overline{g^*} d\mu$$

defines a continuous linear operator on $C(M_A)$. Here f^* denotes the weak*-limit of $(f(r\xi))_{0<r<1}$ relatively μ. See Theorem X:8, XI:5 and Lemma XI:3.

So by the Riesz representation theorem, there is a regular Borel measure $\hat{\mu}$ on M_A so that

$$L(\varphi) = \int \varphi d\hat{\mu}, \quad \forall \varphi \in C(M_A).$$

In particular, $L(\hat{f})=\int f*d\mu=f(0)=\tilde{0}(f)$ so $\hat{\mu}\in M_{\tilde{0}}$.

On the other hand, if $\hat{\mu}\in M_{\tilde{0}}$, consider for $f,g\in A(B)$

$$f\bar{g} \mapsto \int \widehat{\hat{f}g}\,d\hat{\mu}.$$

Then, this determines a continuous linear functional on $C(S)$, $S=\partial B$. Again, since $f\to\int\hat{f}d\hat{\mu}=f(0)=\tilde{0}(f)$, Riesz representation theorem gives a measure $\mu\in M_0$ such that

$$\int f\bar{g}\,d\mu = \int\widehat{\hat{f}g}\,d\hat{\mu}, \quad \forall f,g\in A(B).$$

When $n=1$, M_0 and $M_{\tilde{0}}$ contains only one element. When $n>1$, this is not so and the following two questions are natural to ask.

1. Given $\hat{\mu}\in M_{\tilde{0}}$, determine $\mu\in M_0$ as above. Is it then true that $\int f*\overline{g*}d\mu=\int\widehat{\hat{f}g}\,d\hat{\mu}$, $\forall f,g\in H^{\infty}(B)$?

2. Given $f\in H^{\infty}(B)$. Is it then true that $\lim_{r\to 1}\hat{f}_r(\xi)=\hat{f}(\xi)$, a.e. $(\hat{\mu})$ for all $\hat{\mu}\in M_{\tilde{0}}$?
   (Here $f_r(\xi)=f(r\xi)$; this would be an analogue of Fatou's theorem extended to M_A).

In view of Henkin´s theorem (Theorem XI:5) 2) would imply 1). But 1) is false - assume for a moment that 1) is true.

Then, by Theorem XI:5

$$0 = \inf_{g \in A(B)} \int |g-f*|^2 d\mu = \inf_{g \in A(B)} \int |\hat{g}-\hat{f}|^2 d\hat{\mu}$$

so Theorem XII:1 would give

$$0 = \sup_{\hat{\mu} \in M_{\tilde{0}}} \ \inf_{g \in A(B)} \int |\hat{g}-\hat{f}|^2 d\hat{\mu} = \inf_{g \in A(B)} \ \sup_{\hat{\mu} \in M_{\tilde{0}}} \int |\hat{g}-\hat{f}|^2 d\hat{\mu} =$$

$$= \inf_{g \in A(B)} \ \sup_{\mu \in M_0} \int |g-f*|^2 d\mu .$$

Note that

$$M_{\tilde{0}} \ni \hat{\mu} \mapsto \int |\hat{g}-\hat{f}| d\hat{\mu}$$

is continuous since $|\hat{f}|$ is continuous on M_A for all $f \in H^\infty(B)$.

This give a contradiction, since there exist $f \in H^\infty(B)$ with

$$\inf_{g \in A(B)} \ \sup_{\mu \in M_0} \int |g-f*| d\mu > 0$$

(cf. Theorem X:6 and the example preceeding it).

Notes and references

For results and references concerning Banach algebras we refer to **J.B. Garnett,** Bounded analytic functions. Academic Press, 1981.

Aspects of Mathematics

English-language subseries (E)

Aspekte der Mathematik

Deutschsprachige Unterreihe (D)

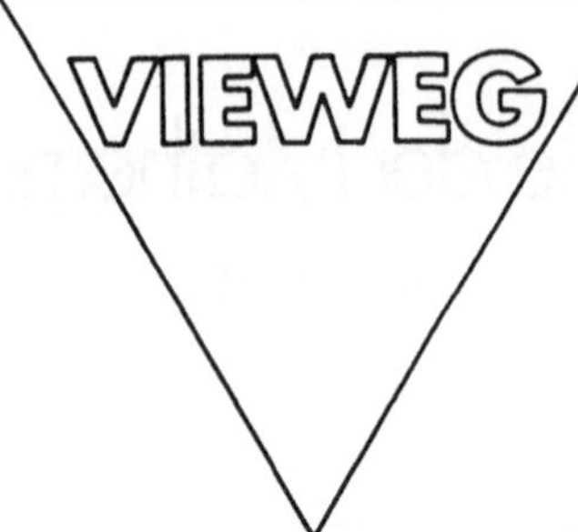

Alan Howard and Pit-Mann Wong (Eds.)

Contributions to Several Complex Variables

In Honour of Wilhelm Stoll. 1986. XII, 352 Seiten. 16,2 x 22,9 cm. (Aspects of Mathematics, Vol. E 9; edited by Klas Diederich.) Softcover.

Contents: Walther L. Bailey, Jr.: Arithmetic Hilbert Modular Functions III – Richard Beals and Nancy K. Stanton: The Heat Equation for the δ-Neumann Problem on Strictly Pseudoconvex Domains – Daniel Burns: Some Examples of the Twistor Construction – Klas Diederich: Complete Kähler Domains. A Survey of some Recent Results – Maria L. Fania and Andrew J. Sommese: On the Minimality of Hyperplane Sections of Gorenstein Threefolds – Hans Grauert: On Meromorphic Equivalence Relations – Alan T. Huckleberry and Wolfgang Richthofer: Recent Developments in Homogeneous CR-Hypersurfaces – Pierre Lelong: Problems of Value Distribution in Complex Analysis for Several Variables – László Lempert: On the Boundary Behavior of Holomorphic Mappings – Robert Molzon: Integral Geometry of the Monge-Ampère Operator – Junjiro Noguchi: Logarithmic Jet Spaces and Extensions of de Franchis' Theorem – Bernard Shiffmann: Remarks on the Nakano Vanishing Theorem – Yum-Tong Siu: Curvature of the Weil-Petersson Metric in the Moduli Space of Compact Kähler-Einstein Manifolds of Negative First Chern Class – Henri Skoda: Extension Problems and Positive Currents in Complex Analysis.

The articles in this volume give an insight into recent research activities in many different branches of complex analysis such as value distribution theory, meromorphic maps, positive currents, complex differential and algebraic geometry, the δ-Neumann problem, homogeneous manifolds, arithmetic Hilbert modular functions and others.